HOMOTOPY
METHODS AND
GLOBAL CONVERGENCE

NATO CONFERENCE SERIES

I	Ecology
II	Systems Science
III	Human Factors
IV	Marine Sciences
V	Air–Sea Interactions
VI	Materials Science

II SYSTEMS SCIENCE

HOMOTOPY METHODS AND GLOBAL CONVERGENCE

Edited by

B. Curtis Eaves
Stanford University
Stanford, California

Floyd J. Gould
University of Chicago
Chicago, Illinois

Heinz-Otto Peitgen
University of Bremen
Bremen, Federal Republic of Germany

and

Michael J. Todd
Cornell University
Ithaca, New York

Published in cooperation with NATO Scientific Affairs Division

PLENUM PRESS · NEW YORK AND LONDON

Library of Congress Cataloging in Publication Data

NATO Advanced Research Institute on Homotopy Methods and Global Convergence
(1981: Porto Cervo, Sardinia) Homotopy methods and global convergence.

(NATO conference series. II, Systems science; v. 13)
Includes bibliographical references and index.
1. Fixed point theory—Congresses. 2. Homotopy theory—Congresses. 3. Convergence—Congresses. I. Eaves, B. Curtis. II. North Atlantic Treaty Organization.
Scientific Affairs Division. III. Title. IV. Series.

QA329.9.N37 1981   514'.24   82-16547

ISBN-13:978-1-4613-3574-0   e-ISBN-13:978-1-4613-3572-6

DOI: 10.1007/978-1-4613-3572-6

Proceedings of a NATO Advanced Research Institute on Homotopy Methods and
Global Convergence, held June 3–6, 1981, in Porto Cervo, Sardinia

© 1983 Plenum Press, New York

Softcover reprint of the hardcover 1st edition 1983

A Division of Plenum Publishing Corporation
233 Spring Street, New York, N.Y. 10013

PREFACE

This Proceedings presents refereed versions of most of the
papers presented at the NATO Advanced Research Institute on
Homotopy Methods and Global Convergence held in Porto Cervo,
Sardinia, June 3-6, 1981. This represents the fourth recent
occurrence of an international conference addressing the common
theme of fixed point computation. The first such conference,
titled "Computing Fixed Points with Applications," was held in the
Department of Mathematical Sciences at Clemson University,
Clemson, South Carolina, June 26-28, 1974 and was sponsored by the
Office of Naval Research and the Office of the Army Research
Center. The second conference, "Symposium on Analysis and
Computation of Fixed Points," was held at the University of
Wisconsin, Madison, May 7-8, 1979, under the sponsorship of the
National Science Foundation, the U.S. Army, and the Mathematics
Research Center of the University of Wisconsin, Madison. The
third conference, titled "Symposium on Fixed Point Algorithms and
Complementarity," was held at the University of Southampton,
Southampton, UK, July 3-5, 1979 and was sponsored by U.N.E.S.C.O.,
European Research Office (London), Department of Mathematics
(University of Southampton), I.B.M. U.K., Ltd., Lloyds Bank, Ltd.,
and the Office of Naval Research (London).

The Advanced Research Institute held in Sardinia was devoted
to the theory and application of modern homotopy methods. The
following topics were stressed: Path-Following Techniques;
Bottom-Line Applications; Global vs. Classical Methods; and State-

v

of-the-Art, Perspectives and Potential. The papers presented were
selected so as to devote more or less uniform attention to these
four areas. In addition, workshop sessions were held on different
days in each of these four subject areas. While the papers repro-
duced herein will serve to memorialize the formal presentations,
the informal presentations and interactions during the workshops,
in spite of their value, and in spite of the efforts of several
members of the organizing committee to edit transcriptions, will
not be documented. The stimulation provided by these interactions
will hopefully be a source of future motivation for the participants and thus, indirectly, will be captured in their work.

A final and somewhat unique feature of this volume is a list
of some computer codes currently in use for implementing the
homotopy method. Descriptions of these codes have been provided
by the originators. Researchers in the field who wish to profit
from the existence of any of these codes may directly contact the
author of the code.

I am indebted to the Systems Science Programme of the
Scientific Affairs Division of NATO for their generous support of
this Institute, to Professor Jean Abadie for his initial encouragement, to Professor Donald Clough for his many suggestions and
thoughtful guidance, to the Organizing Committee, consisting of
Professors Michael J. Todd, James Yorke, Heinz-Otto Peitgen, and
Herbert E. Scarf, to all of the participants for their lively
contributions, and finally to Ms. Maggie Newman for her grudging
devotion and her many and varied contributions to the success of
the Institute as well as the production of this Proceedings.

F. J. Gould, Director
Chicago, Illinois

CONTENTS

PIECEWISE SMOOTH HOMOTOPIES

J. C. Alexander[1]
Department of Mathematics and
Institute for Physical Science and Technology
University of Maryland
College Park, MD 20742 U.S.A.

T.-Y. Li[1,2]
Department of Mathematics
Michigan State University
East Lansing, MI 48823 U.S.A.

J. A. Yorke[2]
Department of Mathematics and
Institute for Physical Science and Technology
University of Maryland
College Park, MD 20742 U.S.A.

1. INTRODUCTION

In [1, 2, 3, 7, 14, 16] there is developed a class of continuation methods for solving nonlinear systems of equations which
have the feature that, under broad topological assumptions which
guarantee the existence of solutions of the system, the methods are
guaranteed with probability one to generate a curve which approaches
arbitrarily close to a solution of the system. In the above papers
it is assumed that the nonlinear system is defined by smooth functions. Piecewise linear techniques are similarly used; see for

---

[1]Supported in part by the National Science Foundation.
[2]Supported in part by Army Research Office grant DAAG-29-80--C-0040.

example [5]. The purpose of this paper is to develop path following
methods for a class of problems including both piecewise linear and
smooth systems of equations. We formulate the method for "piecewise smooth functions" on a "piecewise smooth domain," and we give
similar guaranteed convergence results.

   The concepts of piecewise smooth manifolds and functions can
be defined in a variety of ways, to fit the problem at hand. In
[1, Appendix 1], we announced preliminary results for the simplest
useful version of these definitions. We adopt that version here.

   As an illustration of the kinds of problems we want to be able
to handle, we let B be the ball in R^n and let $f : B \to B$ be
piecewise smooth in the sense defined in the next section. (In
particular we assume f is continuous.) Following the homotopy
approach formally we choose $z \in B$ and write the homotopy

$$F_z(x,t) = (1-t)z + tf(x) - x$$

where $t \in [0,1]$. The zeroes of $F_z(1,x)$ are the fixed points of
f while z is the unique zero of $F_z(0,x)$. When f is smooth
(C^2), it is shown in [3] that for almost every $z \in B$ a smooth
path in $B \times [0,1]$ leads from $(0,z)$ to at least one zero at
t = 1. The objective of this paper is to develop a corresponding
theory which permits f to be piecewise smooth and to show there
is a piecewise smooth path of zeroes of F_z that leads to a fixed
point (or possibly to a larger set of fixed points) of f. The
facts about the paths for F_z follow from the general theory we
develop here, and we develop only enough theory for us to handle applications. We give applications to show how the piecewise
smooth formulation can be used, and these are discussed in more detail. First we consider the nonlinear complementarity problem.
We put it in our context and prove an existence result. The continuation method we develop is a nonlinear form of Lemke's algorithm.
Second we consider nonlinear constrained optimization.

2.   THE PIECEWISE SMOOTH FORMULATION

We set some notation. Let $<x,y>$ denote the inner product in n-dimensional Euclidean space R^n. For $U \subset R^n$ an open set, we speak of smooth mappings $F : U \to R^m$ where "smooth" means C^k for k large ($k \geq 2$ is usually sufficient). If F is smooth, let $DF(x)$, for $x \in U$, denote the $m \times n$ matrix of first partial derivatives of F. Let $y \in R^p$, $z \in R^q$, $p + q = n$. We denote by F_z the map from a domain of R^p to R^m defined by holding z fixed in (y,z). We let $D_y F = DF_z$ be the derivative of F with respect to the y variables. Let $I = [0,1]$.

For convenience we recall the development in [1]. Let M be an n-dimensional topological manifold. Let $U_1, \ldots, U_I$ be a finite open cover of M. Each is to have a smooth structure compatible with its structure as an open topological submanifold of M (a local __smoothing__ of M). Suppose for each $i \in \{1, \ldots, I\}$ there are defined smooth functions

$$\psi_{i,1}, \ldots, \psi_{i,J} : U_i \to R$$

where $J = J(i)$, which satisfy the following transversality conditions:

i) 0 is a regular value of each $\psi_{i,j} : U_i \to R$, i.e., rank $(D\psi_{i,j}(x)) = 1$ if $\psi_{i,j}(x) = 0$;

ii) 0 is a regular value of each $\psi_{i,j} \times \psi_{i,k} : U_i \to R \times R$ if $j \neq k$; i.e., rank $\begin{pmatrix} D\psi_{i,j}(x) \\ D\psi_{i,k}(x) \end{pmatrix} = 2$ if $\psi_{i,j}(x) = \psi_{i,k}(x) = 0$.

Thus the $\psi_{i,j}^{-1}(0)$ are codimension 1 submanifolds of M which meet pairwise transversally. For each i, let

$$V_i = \{x \in M : \psi_{i,j}(x) > 0 \text{ for } j = 1, \ldots, J(i)\}.$$

We require that $V_i \cap V_{i'} = \phi$ if $i \neq i'$, that each $\overline{V}_i \subset U_i$ and that $M = \cup \overline{V}_i$. Such a collection of data we call a __piecewise smooth decomposition__ of M. The V_i are to be pieces on which things are smooth. The <u>edges</u> of M_i consist of those x such that

some $\psi_{i,j}(x) = 0$. The <u>corners</u> of M_i are those sets where for

some i

$$\psi_{i,j}(x) \;=\; \psi_{i,k}(x) \;=\; 0$$

for some $j \neq k$. The V_i themselves are the <u>regions</u> or <u>faces</u> of

the decomposition. Note that an edge point that is not a corner is

either on the boundary of M or is an edge point for precisely two

faces.

A continuous map $f : M \to R^n$ is called <u>piecewise smooth</u> if

there is a collection of smooth functions $f_i : U_i \to R^n$ for i =

1,...,I such that $f|V_i = f_i$. Intuitively, f is smooth on each

face V_i but can have sharp turns at the edges. Let $\pi : M \times I \to M$

be the projection. We define a piecewise smooth decomposition of

$M \times I$ by letting $\tilde{U}_i = \pi^{-1}U_i$, $\tilde{\psi}_{i,j} = \psi_{i,j}\pi$. Thus $M \times I$ is decomposed

by cylinders $V_i \times I$. Let A (the <u>parameter</u> <u>manifold</u>) be a smooth

manifold. Now suppose $\phi : A \times M \to R^n$ is a smooth function and let $f_\alpha(x)$

$= \phi(\alpha,x)$ for each $\alpha \in A$. Let $F_\alpha : M \times I \to R^n$ be defined by

$$F_\alpha(x,t) \;=\; (1-t)f_\alpha(x) + tf(x).$$

Similarly define $F_{i,\alpha} : U_i \times I \to R^n$ by

$$F_{i,\alpha}(x,t) \;=\; (1-t)f_\alpha(x) + tf_i(x).$$

We say $F_\alpha^{-1}(0)$ is <u>transverse</u> <u>to</u> <u>the</u> <u>decomposition</u> <u>of</u> $M \times I$ if

for $t < 1$

t-0)   0 is a regular value of each $F_{i,\alpha} : U_i \times I \to R^n$

t-i)   0 is a regular value of each

$$F_{i,\alpha} \times \tilde{\psi}_{i,j} : U_i \times I \to R^n \times R$$

t-ii)   0 is a regular value of each

$$F_{i,\alpha} \times \tilde{\psi}_{i,j} \times \tilde{\psi}_{i,k} : U_i \times I \to R^n \times R \times R \quad \text{for} \quad j \neq k.$$

Condition t-0) implies that each $F_{i,\alpha}^{-1}(0)$ is a smooth curve and thus $F_\alpha^{-1}(0)$ is a piecewise smooth curve. Condition t-i) implies that each $F_{i,\alpha}^{-1}(0)$ intersects each edge transversally. Condition t-ii) implies $F_{i,\alpha}^{-1}(0)$ contains no corner points.

Recall that $\phi : A \times M \to R^n$ is <u>sufficient</u> if rank $D_\alpha \phi(\alpha,x) = n$ for all (α,x). The following is the main result. A proof is sketched in [1]. It is standard, except that extra care has to be taken at the edges and corners.

<u>Theorem 1.</u> If $\phi : A \times M \to R^n$ <u>is sufficient, the set</u> $F_\alpha^{-1}(0)$ <u>is transverse to the decomposition with probability</u> 1 <u>in</u> α (<u>i.e., for a full residual set of</u> α).

Thus with probability 1, $C = C_\alpha = F_\alpha^{-1}(0)$ is a piecewise smooth curve intersecting edges transversally and corners not at all. In applications, one follows C from the $t = 0$ level to a solution of $f(x) = 0$ at $t = 1$.

3. EXAMPLES

Since most applications concern maps $f : R^n \to R^n$, the basic manifold M is R^n. We present the orthant piecewise smooth decomposition. It is an easy one to visualize and is used in the application of the next section.

Let i range over the 2^n subsets of the integers $1,\ldots,n$. For each i, let $U_i = R^n$. Let each $J(i) = n$. Let

$$\psi_{i,j}(x_1,x_2,\ldots,x_n) = \begin{cases} x_j & j \in i, \\ -x_j & j \notin i. \end{cases}$$

Note that

$$D\psi_{i,j}(x_1,\ldots,x_n) = (0,\ldots,0,\pm 1,\ldots,0)$$

(here the ± 1 is in the j<u>th</u> position)

and

$$D(\psi_{i,j} \times \psi_{i,k})(x_1,\ldots,x_n) = \begin{pmatrix} 0,\ldots,0,\pm1,0,\ldots,0,0,0,\ldots,0 \\ 0,\ldots,0,0,0,\ldots,0,\pm1,0,\ldots,0 \end{pmatrix}$$

(here the ±1 are in the jth kth columns).

Thus conditions (i), (ii) are satisfied and we have defined a piece-wise smooth decomposition of R^n, the orthant decomposition. Note

$$V_i = \{(x_1,\ldots,x_n) : x_j > 0 \text{ if } j \in i \text{ and } x_j < 0 \text{ if } j \notin i\}.$$

The edges are the coordinate hyperplanes and the corners are where two or more coordinate hyperplanes intersect. If a curve C is transverse to the orthant decomposition (or its product with I), at each edge exactly one coordinate of the position vector of C changes sign.

The orthant decomposition is of course actually piecewise linear, and it could be hoped a simpler theory could be used. However, we are considering functions that are not piecewise linear, and it seems the full concept of piecewise smooth manifolds is needed. Of course, an actual implementation might benefit from the simplicity of the orthant decomposition.

The definition is flexible enough to accommodate manifolds with "sharp bends". For example, the surface of a cube in three-dimensional space is topologically equivalent to the surface of a sphere, however, the surface of the cube has a different piecewise differentiable structure.

To illustrate in more detail, we consider the simplest case. Suppose M is the union of the non-negative parts of the two-dimensional space:

$$M = \{(x,y) : x \geq 0 \text{ or } y \geq 0\}.$$

We put the natural piecewise differentiable structure on M. Let $U_1 = U_2 = M$. Both U_i are to be smoothly like R^1 via the homeo-morphism $h : U_i \to R^1$:

$$h(x,0) = x,$$

$$h(0,y) = -y.$$

Then the functions $\psi_1 : U_1 \to R^1$, $\psi_2 : U_2 \to R^1$:

$$\psi_1(x,0) = x, \qquad \psi_2(x,o) = -x,$$

$$\psi_1(0,y) = -y, \qquad \psi_2(0,y) = y$$

are smooth. The required conditions on the ψ_i are satisfied. The faces V_1, V_2 are respectively the positive x- and y-axes. The sole edge is the origin.

A function on M defines a piecewise smooth function if it is smooth on each axis (using one-sided derivatives at the origin). Each of the two pieces of the function—one on each axis—can be extended smoothly to all of the U_i. Note that the U_i are technical artifacts. For manifolds with sharp bends, the key requirement for piecewise smooth functions is that good one-sided derivatives exist at the edges. The example we present of constrained optimization is a more complicated case in point.

4. NONLINEAR COMPLEMENTARITY

Let $g : R^n \to R^n$ be smooth. The complementarity problem is to find an $x \in R^n$ such that

$$x \geq 0, \qquad g(x) \geq 0, \qquad <x,g(x)> = 0.$$

If $s \in R$, let $s^+ = \max(0,s)$, $s^- = \min(0,s)$ and if $x = (x_1,\ldots,x_n) \in R^n$, let $x^{\pm} = (x_1^{\pm},\ldots,x_n^{\pm})$. We write $x < y$ if each $x_i < y_i$, $i = 1,\ldots,n$. Let $R_{\pm}^n = \{x : x^{\pm} = x\}$. Consider the map $f : R^n \to R^n$ defined by

$$f(x) = x^- + g(x^+).$$

If $f(x_0) = 0$ for some $x_0 \in R^n$, then x_0^+ is a solution of the complementarity problem associated with f. For $f(x_0) = 0$ implies

$$0 = \langle x_0^+, f(x_0) \rangle = \langle x_0^+, x_0^- \rangle + \langle x_0^+, g(x_0^+) \rangle = \langle x_0^+, g(x_0^+) \rangle$$

and $g(x_0^+) = -x_0^- \geq 0$.

Moreover, note that $f(x)$ is piecewise smooth with respect to the orthant subdivision of R^n. So we set $\phi : R^n \times R^n \to R^n$ as $\phi(z,x) = x^- - z$. Thus the homotopy

$$F_z(x,t) = tg(x^+) + x^- - (1-t)z.$$

Since $-D_z\phi$ is the identity matrix, ϕ is sufficient. Thus for almost all z, $F_z^{-1}(0)$ is a piecewise smooth curve transverse to the decomposition. Let $C = C_z$ be the component continuing $(z,0)$. To ensure the curve converges to a solution, we use a variation of a condition of Eaves [4]. Related conditions are in [5, 8, 11, 12, 17].

Eaves' condition is: for a non-empty open set Ω of $z > 0$, there is a bounded set $U(z)$ which intersects any closed unbounded connected set in R_+^n which contains 0 (i.e., $U(z)$ <u>separates</u> 0 from ∞ in R_+^n), and such that for each $x \in U(z)$ there is $w \in R_+^n$ such that

$$\langle x-w, z \rangle > 0 \quad \text{and} \quad \langle x-w, g(x) \rangle \geq 0.$$

Let $-\Omega = \{z \in R^n : -z \in \Omega\}$.

Theorem 2. <u>Suppose Eaves' condition holds. For almost all</u> $z \in -\Omega$, <u>the curve</u> $C = C_z$ <u>is a piecewise smooth curve containing</u> $(z,0) \in R^n \times I$ <u>and eventually remaining arbitrarily close to the</u> (<u>non-empty</u>) <u>set</u> $\{(f^{-1}(0),1)\} \subset R^n \times I$.

Proof. We need only prove the last statement. First note that if $(x,0) \in C$, then $x = z$. This is because $x^- = z$, thus all components of x^- are non-zero and $x^+ = 0$. Thus C cannot return to the $t = 0$ level. Next we show C is bounded. If not, there exist (x,t) on C with either $|x^+| \to \infty$ or $|x^-| \to \infty$. Suppose

$|x^+|$ remains bounded. Then $x^- = -tg(x^+) + (1-t)z$ also remains bounded. Thus $|x^+| \to \infty$. Consider the set $U(-z)$ from Eaves' condition. It is bounded, say it is contained in the ball of radius $\frac{1}{2}$ P. The set

$$C^+ = \{x^+ : (x,t) \in C\}$$

is unbounded and connected in R_+^n and contains 0. There is some $t_0 < 1$ such that $(x_0,t_0) \in C$ with $|x_0^+| > P$. Consider the set

$$C_0^+ = \{x^+ : (x,t) \in C, t \leq t_0\} \cup \{x^+ \in R_+^n : |x^+| \geq P\}.$$

It is a closed, unbounded, connected set in R_+^n which contains 0. By Eaves' condition, the set $U(-z)$ intersects C_0^+. So, there is $(x,t) \in C$ such that $x^+ \in U(-z)$. Let $w \geq 0$ be as in Eaves' condition. Then

$$0 = \langle F_z(x,t), x^+ - w\rangle = t\langle g(x^+), x^+ - w\rangle + \langle x^-, x^+ - w\rangle - (1-t)\langle z, x^+ - w\rangle$$

$$= t\langle g(x^+), x^+ - w\rangle - \langle x^-, w\rangle - (1-t)\langle z, x^+ - w\rangle$$

$$> 0$$

since $t < 1$. Thus C must remain bounded. The rest of the argument is standard (e.g. [7], Theorem 2.2).

5. CONSTRAINED OPTIMIZATION

Let $f(x)$ and $g_i(x)$, $1 \leq i \leq m$ be smooth real valued functions on R^n. We consider the nonlinear programming problem:

minimize $f(x)$ with $g_i(x) \leq 0$.

As is well known [10] if x^* is a solution of this problem, there exist numbers λ_j, $1 \leq j \leq m$ (Lagrange multipliers) such that

$$Df(x^*) + \sum_1^m \lambda_j Dg_j(x^*) = 0$$

(5.1)

$$g_j(x^*) \leq 0, \quad \lambda_j \geq 0, \quad \lambda_j g_j(x^*) = 0, \quad 1 \leq j \leq m.$$

This formulation is the one actually solved.   To show   x*   is actually a minimum usually requires further checking.   Homotopy methods for solving such non-linear problems have been developed in
[9, 13, 15].

Equations (5.1) look something like a complementarity problem.
That observation leads to the following homotopy method.   Consider
the mapping $G : R^{n+m} \to R^{n+m}$ defined by

$$
G(x,\lambda) \quad = \quad \begin{pmatrix} Df(x) \ + \ \sum_{1}^{m} \lambda_j^+ Dg_j(x) \\ \\ \lambda_i^- - g_i(x) \end{pmatrix}_{1 \le i \le m.}
$$

Then (5.1) holds if $G(x,\lambda) = 0$.   Moreover, G is piecewise
smooth with respect to the orthant decomposition of R^{n+m}.   Let

$$
\Omega \quad = \quad \{x \in R^n : g_i(x) < 0 \quad i = 1,\ldots,m\}.
$$

Assume $\Omega \ne \phi$.   Consider the homotopy $F : \bar{\Omega} \times R^{n+m} \times I \to R^{n+m}$ defined by

$$
F(z,x,\lambda,t) \quad = \quad \begin{pmatrix} t[Df(x) + \sum_{1}^{m} \lambda_j^+ Dg_j(x)] + (1-t)(x-z) \\ \\ \lambda_i^- - g_i(x) \end{pmatrix}_{1 \le i \le m.}
$$

This would seem to give a reasonable homotopy algorithm.   However,
as is, it cannot be proved to be by the general theory.   The reason
is that DF_z never has full rank; the last   m   rows are always
zero.   Indeed   z   ranges over an   n-dimensional set and the range
is   (m+n)-dimensional.   The problem is that the constraints
$\lambda_i^- - g_i(x) = 0$ are independent of   z.   For certain innocuous g_i
(e.g., some $g_i(x) = x_i$),   there are families of curve with segments totally contained within edges of the orthant decomposition.

We overcome this problem by reinterpreting the homotopy problem on an n-dimensional manifold, although an implementation can use the above F directly.

We need a standard condition on the g_i. For any $x \in \bar{\Omega}$, let $I(x) = \{i : g_i(x) = 0\}$. We assume: for any x, the vectors $(Dg_i(x))_{i \in I(x)}$ are linearly independent. Let

$$M = \{(x,\lambda) : g_i(x) \leq 0, \quad \lambda_i \geq 0, \quad \lambda_i g_i(x) = 0, \quad 1 \leq i \leq m\}.$$

It is straightforward, using the above condition, to put a piecewise smooth structure on M so that the following sets V_I are the re-gions. Let $I \subset \{1,\ldots,m\}$ be an arbitrary subset (possibly empty). Then

$$V_I = \{(x,\lambda) : g_i(x) = 0 \text{ for } i \in I, \quad \lambda_j = 0 \text{ for } j \notin I\}.$$

Thus the manifold M incorporates the constraints $g_i(x) \leq 0$ in its definition. The homotopy F on M becomes

$$F(z,x,\lambda,t) = t[Df(x) + \sum_1^m \lambda_i Dg_i(x)] + (1-t)(x-z).$$

This homotopy has the standard form with $\phi(x,z) = x - z$. It is triv-ial that ϕ is sufficient; thus for almost all $z \in \Omega$, the set $F_z^{-1}(0)$ is a piecewise smooth curve. It is precisely the curve ob-tained from the earlier homotopy; we now know for almost all z it is well-behaved at edges and misses corners.

To get a convergence result, we impose rather standard condi-tions. Assume $\bar{\Omega}$ is convex. Assume there is a compact set $A \subset R^n$ such that for $x \notin A$,

$$<x, Df(x)> > 0, \qquad <x, Dg_i(x)> > 0, \qquad i = 1,\ldots,m.$$

Theorem 3. Under these assumptions and the assumption on the $Dg_i(x)$, for almost all $z \in \Omega$, the set $C = F_z^{-1}(0)$ is a piecewise smooth curve in $M \times I$ containing $(z,0,0)$ which eventually remains arbitrarily close to the (non-empty) set of $(x^*, \lambda, 1)$ satisfying (5.1).

Proof. Only the last statement requires proof. If $t = 0$, $(z, x, \lambda, t) \in C$, then $x = z$. Thus the curve cannot return to the $t = 0$ level. Suppose $|x| \to \infty$ on C. Then for $x \notin A$, $t < 1$,

$$0 = <x, F(z, x, \lambda, t)> = t<x, Df(x)> + t \sum_1^m \lambda_i <x, Dg_i(x)> + (1-t)<x, x-z>.$$

Since $|<x, x-z>| \geq |x|^2 - |x||z|$ and z is fixed, this is a contradiction. Thus suppose $|\lambda| \to \infty$ on C. Because the $(Dg_i(x))_{i \in I}$ are linearly independent, $t \to 0$. If $t|\lambda| \to 0$, then $x - z = 0$, which we have already excluded. Thus

$$\frac{Df(x)}{|\lambda|} + \sum_1^m \frac{\lambda_i}{|\lambda|} Dg_i(x) + \frac{1-t}{t|\lambda|}(x-z) = 0.$$

We extract a sequence of points on C such that

$$x \to x^* \in \bar{\Omega}, \qquad \frac{\lambda}{|\lambda|} \to \lambda^*, \qquad \frac{1-t}{t|\lambda|} \to \alpha \geq 0.$$

Then

$$\sum_1^m \lambda_i^* Dg_i(x^*) + \alpha(x^* - z) = 0.$$

Since x^* is on the boundary of $\bar{\Omega}$ and $z \in \bar{\Omega}$, this equation violates the convexity of $\bar{\Omega}$. Thus $|\lambda|$ remains bounded.

The rest of the proof is standard.

Remarks. 1. We defined M directly. However the interested reader can, using the implicit function theorem in a piecewise manner, define piecewise smooth submanifolds of other manifolds, and realize M as a piecewise smooth submanifold of the orthant decomposition on R^{n+m}.

2. If f and the g_i are convex functions, the value of t on the curve C of Theorem 3 is strictly increasing. Thus t can be used as a parameter for the curve.

Acknowledgment. We would like to thank R. B. Kellogg for several comments and helpful suggestions. We are also indebted to the referees, who read the paper most carefully and made a number of pertinent suggestions.

REFERENCES

1. J. C. Alexander, "The topological theory of an imbedding method," in Continuation Methods, Hj. Wacker, ed., Academic Press, New York (1978), 37-68.

2. J. C. Alexander and J. A. Yorke, "The homotopy continuation method: numerically implementable topological procedures," Trans. Amer. Math. Soc., 242 (1978), 271-284.

3. S. N. Chow, J. Mallet-Paret and J. A. Yorke, "Finding zeros of maps: homotopy methods that are constructive with probability one," Math. Comp. 32 (1978), 887-899.

4. B. C. Eaves, "On the basic theorem of complementarity," Math. Prog. 1 (1971), 68-75.

5. B. C. Eaves and H. Scarf, "The solution of systems of piecewise linear equations," Math. Oper. Res., 1 (1976), 1-27.

6. S. Karamardian, "The complementarity problem," Math. Prog. 2 (1972), 107-129.

7. R. B. Kellogg, T.-Y. Li, and J. A. Yorke, "A constructive proof of the Brouwer fixed-point theorem and computational results," SIAM J. Numer. Anal. 13 (1976), 473-483.

8. M. Kojima, "A unification of the existence theorems of the nonlinear complementarity problem," Math. Prog. 9 (1975), 257-277.

9. _____, "A complementary pivoting approach to parametric nonlinear programming," Math. Op. Research 4 (1979), 464-477.

10. D. G. Luenberger, Introduction to linear and nonlinear programming, Addison-Wesley, Reading, Mass. (1973).

11. J. J. Moré, "Coercivity conditions in nonlinear complementarity problems," SIAM Review 16 (1974), 1-16.

12. _____, "Classes of functions and feasibility conditions in non-

linear complementarity problems, "Math. Prog. 6 (1974), 327-338.

13. R. Saigal, "The fixed point approach to nonlinear programming," Mathematical Programming Study 10 (1979), 142-157.

14. S. Smale, "Convergent process of price adjustment and global Newton methods," J. Math. Econom. 3 (1976), 107-120.

15. M. J. Todd, "A quadratically-convergent fixed-point algorithm for economic equilibria and linearly constrained optimization," Math. Prog. 18 (1980), 111-126.

16. L. T. Watson, "An algorithm that is globally convergent with probability one for a class of nonlinear two-point boundary value problems," SIAM J. Numer. Anal. 16 (1979), 394-401.

17. _____, "Solving the nonlinear complementarity problem by a homotopy method," SIAM J. Control Optimization, 17 (1979), 36-46.

GLOBAL CONVERGENCE RATES OF PIECEWISE-LINEAR CONTINUATION METHODS:

A PROBABILISTIC APPROACH

J. C. Alexander[1]
Department of Mathematics and
Institute for Physical Science and Technology
University of Maryland
College Park, MD 20742 U.S.A

E. V. Slud
Department of Mathematics and
Institute for Physical Science and Technology
University of Maryland
College Park, MD 20742 U.S.A

1. INTRODUCTION

Part of the theory of any method of numerical computation consists of results on convergence—under what conditions can convergence be guaranteed and at what rate. For continuation methods, the technique of proving convergence is fairly well understood. The usual results relate—explicitly or implicitly—to topological invariants such as degree. Results about the rate of convergence are much more difficult. To a large extent, this is because continuation methods are global. Just as global topological invariant are needed to guarantee convergence, global studies of rates of convergence are needed. Such global studies promise to be difficul and are only now beginning—for example, Smale's recent work on finding roots of polynomials [5].

---

[1] Partially supported by NSF.

In this paper, we begin the study of rates of convergence for piecewise linear homotopy methods. We cannot claim any definitive results. It is not even clear what form final results should take. Rather we present here some methods that can be brought to bear on the problem, and we report some preliminary results.

There have been a few systematic experimental studies of rates of convergence, e.g., by Saigal [4] and [unpublished] and Todd [unpublished]. Our preliminary theory is not ready to be compared quantitatively with experimental results; rather the experimental results indicate in what directions the theory needs further development. It is hoped that in the not-too-distant future, the theory will be mature enough to complement the experimental results.

We assume the reader understands the basic mechanisms of piecewise-linear continuation methods. We use Todd's monograph [6] as a basic general reference. We work in the following general context. We have a continuous map $f : \mathbb{R}^d \to \mathbb{R}^d$ and we wish to solve $f(x) = 0$. We introduce a homotopy parameter t and consider some continuous map $F : \mathbb{R}^d \times [0,1] \to \mathbb{R}^d$. We assume

$$F(x,0) = f_0(x), \quad F(x,1) = f(x)$$

where $f_0 : \mathbb{R}^d \to \mathbb{R}^d$ is some "initial" known function. The space $\mathbb{R}^d \times [0,1] \subset \mathbb{R}^{d+1}$ is triangulated or otherwise decomposed in a piecewise-linear manner. The map F is approximated by a piecewise-linear map $\widetilde{F}$, and the piecewise-linear path $\widetilde{F}^{-1}(0)$ is numerically "followed" from the $t = 0$ level to the $t = 1$ level. Success is guaranteed by some kind of global convergence result. In practice, the algorithm pivots across faces of the decomposition using, via a labelling, the values of F at vertices. Moreover the t-axis is usually divided into subintervals, say

$$0 = t_0 < t_1 < \ldots < t_n = 1.$$

This provides a natural decompositon of the analysis; each subinterval is analyzed separately and then the separate results are combined. In what follows we distinguish between second generation algorithms (Merrill type, with no retrogression in t) and third generation algorithms (with retrogression allowed).

Other piecewise-linear processes, for example those that suppress the t-direction, can be translated into the above formulation.

A worst-case deterministic analysis is not called for. Todd and others have constructed examples where the pivoting process passes, at least locally, through every simplex of a triangulation. Some kind of probabilistic analysis is more appropriate. Various "measures of efficiency" have been proposed by various workers [3, 4, 6]. Consider for example, Todd's average directional density. It measures the expected number of intersection of a "random" straight line with faces of the decomposition per unit length of the line. B. C. Eaves and J. A. Yorke announced at Sardegna that, using geometric integration theory, they can interpret this measure directly in terms of the geometry of the triangulation. Such a measure is a good probabilistic local measure; it measures the rate at which the pivoting process is proceeding. However, a global measure of rate of convergence must take the total length of the path into account.

We propose a completely different approach. Rather than consider a random path, we consider a random pivoting process. The stochastic models we use are Markov chains (see [1]), which means that we make the simplifying assumption that the pivot from a given face upon entering a given cell of the triangulation does not depend on which previous pivots have been taken. Our technique is to develop global convergence rates from expected first-passage times (for this terminology, see [1, Chap. 15]) of the random-pivot Markov chain. The Markov assumption is the simplest possible assumption about the transition probabilities of the local pivoting process, and are likely too simple to completely describe the stochastic behavior of the pivoting process. We make these assumptions largely for lack of better information. Unfortunately, it seems there have been no investigations into the local stochastic bahavior of pivoting. This appears to be a fruitful area for both experimental and theoretical ressearch.

2. MACRO-ANALYSIS

We break the analysis into two parts. The t-axis has been subdivided into n intervals. Within each interval the x-space $\mathbb{R}^d$ is decomposed piecewise-linearly. The analysis is correspondingly divided into an analysis in the x-direction and in the t-direction. We call these respectively, the <u>micro-</u> and <u>macro-analysis</u>. The micro-analysis depends directly on the details of the triangulation and the local pivoting process. The macro-analysis on the other hand, is of a quite general nature; the resulting formulae involve several parameters. The values of the parameters depend on the results of the micro-analysis. In this section we set the definitions and develop the formulae of the macro-analysis. The analysis for second-generation and third-generation algorithms are slightly different; the latter is slightly more general.

Thus we consider a finite number of t-values

$$0 = t_0 < t_1 < \ldots < t_n = 1.$$

We call t_r the <u>rth level</u>, and consider going from t_r to $t_{r\pm 1}$ a <u>macro-step</u>. A macro-step consists of a number of <u>micro-steps</u>, each of which is one pivot, each of which takes one unit of time. Thus time and number of pivots are equal.

When the process crosses a level r, it crosses either in the forwards (+) or backwards (-) direction. The macro state space S then consists of the set

$$\{r+, \ r- : r = 0, \ldots, \ n\}$$

where r signifies crossing the rth level and ± indicates the direction. The process starts at 0+ and ends when it arrives at n+. The waiting time for this to occur—also called a "first-

passage time" in the terminology of Markov chains—is the object of
our interest.

There are two types of transitions we consider in this state
space. The first is relevant for third-generation algorithms, where
both forward and backward motion is permitted. The second is relevant for second-generation algorithms and for the initial level of
third-generation algorithms, that is, whenever reverse motion across
a level is prohibited.

I. From state r+ [respectively r-], the process moves
either to (r+1)+ [(r-1)-] with probability p_+ [p_-] or to r-
[r+] with probability $q_+ = 1 - p_+$ [$q_- = 1 - p_-$]. Given that the
process moves from r+ to (r+1)+ [from r- to (r-1)-], the
time required for the transition step is random with expectation
and second moment s_+, g_+ [respectively s_-, g_-]. Given that
the process moves from r+ to r- [from r- to r+], the transition step requires a random time with expectation and second
moment t_+, h_+ [t_-, h_-].

II. From state r+, the process moves to (r+1)+ with probability 1, with expected time $\bar{s}$, second moment $\bar{g}$.

Remarks. 1. For a second-generation algorithm, only type II
transitions are permitted, and the states r- are never reached.

2. The p_+, p_-, s_+, s_-, etc., are parameters which are
determined by the micro-analysis.

3. The parameters $s_\pm$, $t_\pm$, $g_\pm$, $h_\pm$ are conditional expectations and moments given forward motion; however $\bar{s}$, $\bar{g}$ are absolute,
not conditional, moments.

4. Such a process is called a semi-Markov process [2, p.207]
because the time taken for the current transiton step depends not
only on the present position but also on the immediate destination.

5. For this paper, we restrict ourselves to the special case
$p_+ = p_-$, $s_+ = s_-$, etc., and drop the + and - subscripts on
parameters.

The variables used in the analysis are the following:

$S_1(r)$ = expected time to reach $n+$, starting at $r+$;

$Q_1(r)$ = expected time to reach $n+$, starting at $r-$;

$S_2(r)$, $Q_2(r)$ = respective second moments.

The average waiting time, i.e., total expected number of pivots, from beginning to end of the process is thus $S_1(0)$, and $S_2(0) - S_1(0)^2$ is the associated variance.

We consider a third-generation fixed-point algorithm, where $f_0(x) = c - x$ for c some point of the domain, so that the pivoting process cannot return to $r = 0$. The initial (0th) level has type II transitions; all other levels have type I transitions. The $S_i(r)$, $Q_i(r)$ satisfy a system of difference equations:

$$
\begin{aligned}
S(0) &= \alpha + S(1) \\
Q(1) &= \gamma + S(1) \\
\text{(2.1)} \quad S(r) &= \lambda S(r+1) + (1-\lambda)Q(r) + \beta_r & 1 \le r \le n - 1 \\
Q(r) &= \lambda Q(r-1) + (1-\lambda)Q(r) + \delta_r & 2 \le r \le n - 1 \\
S(n) &= 0
\end{aligned}
$$

In particular, $S_1(r)$, $Q_1(r)$ satisfy (2.1) with

$$
\begin{aligned}
\lambda &= p & \gamma &= t, \\
\beta_r &= \delta_r = ps + qt, & \alpha &= \bar{s};
\end{aligned}
$$

and $S_2(r)$, $Q_2(r)$ satisfy (2.1) with

$\alpha = \bar{s} + 2sS_1(1)$, $\quad \beta_r = B_r \equiv 2[psS_1(r+1) + gtQ_1(r)] + (pg + qh)$,

$\gamma = h + 2tS_1(1)$, $\quad \delta_r = D_r \equiv 2[psQ_1(r-1) + qtS_1(r)] + (pg + qh)$,

$\lambda = p$, $\qquad\qquad$ (but $D_1 \equiv 0$).

For example consider the formula for $S_1(n)$:

$$S_1(r) = pS_1(r+1) + qQ_1(r) + ps + qt.$$

From $r+$, the process moves either to $(r+1)+$ or $r-$. In the first case, which has probability p, the expected waiting time to reach $n+$ is $(S_1(r+1)+s)$. In the second case, with probability q, the expected waiting time is $(Q_1(r)+t)$. Thus the formula. Note that $ps+qt$ is the expected number of micro-steps in one macro-step.

The general solution for system (2.1) (as the reader can verify by substitution) is (with the convention $\delta_1 \equiv 0$)

$$S(r) = (n-r) \; \gamma \frac{1-\lambda}{\lambda} + \frac{1-\lambda}{\lambda} \; (n-r) \sum_{k=1}^{r-1} (\beta_k + \delta_k)$$

$$(2.2) \qquad + \sum_{k=r}^{n-1} [\beta_k + \frac{1-\lambda}{\lambda} (n-k)(\beta_k + \delta_k)] \qquad 1 \le r \le n-1$$

$$Q(r) = S(r) + \gamma - \beta_r + \sum_{k=1}^{r} (\beta_k + \delta_k)$$

In particular

$$(2.3) \quad S_1(0) = \bar{s} + (n-1)[ps + qt + q(t-s)] + n(n-1)[\frac{q}{p}(ps+qt)].$$

$$(2.4) \quad S_2(0) = \bar{g} + 2sS_1(1) + (n-1)\frac{q}{p}[h + 2tS_1(1)]$$

$$+ \sum_{r=1}^{n-1} [B_r + \frac{q}{p}(B_r + D_r)(n-r)].$$

It is instructive to consider lead terms. The lead term in n in (2.3) is

$$(2.5) \qquad\qquad\qquad S_1(0) \sim \frac{q}{p}(ps + qt)n^2.$$

As a result of the micro-analysis, the parameters p, q, s, t depend on the dimension d, and (2.5) is also the lead term in d. Assuming, as is the case in our examples, that the summation terms dominate in (2.4), the lead term is

(2.6) $$S_2(0) \sim \frac{5}{3}\left(\frac{q}{p}\right)^2 (ps + qt)^2 n^4.$$

Consider now the equivalent second-qeneration algorithm, in which only type II transitions are permitted. There are n stochastically independent macro-steps, whose expectations and variances add to give the expectation and variance of the total waiting time to reach n+:

(2.7) $$S_1(0) \;=\; n\bar{s}.$$

(2.8) $$S_2(0) - S_1(0)^2 \;=\; n(\bar{g} - \bar{s}^2).$$

This observation is exactly analogous to the discussion of "ladder points" and "ladder variables" in [1, Chapter 13] for one-dimensional random walks.

Remark. Note that for n large, the variance (2.5) for third generation algorithms is about the square of the expectation (2.6). Thus the standard deviation is of the same order as the expectation. On the other hand, for second generation algorithms, since the macro-steps are stochastically independent, a law of large numbers applies (see [1, Chap. 10]): the standard deviation is of a lower order than the expectation, and the waiting time from 0+ to n+ is asymptotically (as n gets large) the same as its expectation.

3.   MICRO-ANALYSIS FOR RECTANGULAR CELLS: AN ILLUSTRATIVE CASE

We consider first a rectangular decomposition R of $\mathbb{R}^d$. That is, $\mathbb{R}^d$ is divided by hyperplanes parallel to the coordinate hyperplanes. The decomposition R is not particularly relevant to applications, but it serves as the simplest example for micro-analysis. Everything can be computed explicitly, and the geometry and probability theory are transparent.

The numerical path-following process exits from a cell by

pivoting out of one of the faces. For a stochastic analysis, each face has a certain probability of being the exit face. These local probabilities underlie the micro-analysis. Here we adopt the assumption of maximal randomness; all faces are equally likely to be exit faces.

Any actual numerical algorithm never reenters a cell it has left. This is an exclusion principle. Modelling such exclusion principles causes severe complications, obscuring the underlying ideas. Moreover, incorporating such exclusions in the model does not change the analysis much; the answers change only in the low order terms. For the decompositons considered in this paper, the authors have done the analysis for models which exclude an immediate return to the cell which has just been left. Since the answers are virtually the same as for a model without exclusions we present computations only for the simpler model.

Between any two levels, r and (r+1) say, the decomposition R is a slab of rectangular d-dimensional solids, of thickness one. Each rectangular solid has $2d + 2$ faces. The path-following process enters the slab at r+, may move laterally through rectangular solids, and eventually exits forward to (r+1)+ or backwards to r-.

Of the faces of each cell, one exits forward to the next level, one exits backward, and 2d exit laterally. Under the assumption of maximal randomness, the waiting time to exit forward or backward has a geometric distribution [1, Chap. 6] with parameter $\frac{2}{2d+2}$ for type I transitions and $\frac{1}{2d+1}$ for type II transitions. By symmetry, for type I transitions

$$p = \frac{1}{2}, \qquad s = t = d + 1,$$

$$g = h = (2d+1)(d+1).$$

For type II transitions

$$\bar{s} = 2d + 1 \qquad \bar{g} = (2d+1)(4d+1).$$

Accordingly, from (2.3),

(3.1) $S_1(0) = (2d+1) + (d+1)(n^2-1).$

From (2.5),

(3.2) $S_1(0) \sim n^2(d+1) \sim dn^2.$

From (2.6),

(3.3) $S_2(0) \sim \frac{5}{3}(d+1)^2 n^4 \sim \frac{5}{3} d^2 n^4.$

For second-generation algorithms, from (2.7), (2.8)

(3.4) $S_1(0) = (2d+1)n \sim 2dn$

(3.5) $S_2(0) - S_1(0) = (2d+1)2dn \sim 4d^2 n.$

Remark.   The interested reader can verify that for a model
that exclude immediate reversals,

$$p = \frac{d+1}{2d+1}, \qquad q = \frac{d}{2d+1},$$

$$s = \frac{2d^2+3d+2}{2d+2},$$

$$t = \frac{2d+3}{2},$$

$$\overline{s} = \frac{4d^2+1}{2d+1}.$$

Thus the asymptotic relations (3.2) - (3.5) are unchanged.

4. MICRO-ANALYSIS OF K_1 AND J_1 TRIANGULATIONS

The pivoting rules given by Todd [6, p. 35] for triangulation
K_1 lead directly via our assumption of maximal randomness to the
following type-I Markov transition mechanism for pivots, defined
on a state space of labels $S = \{(I,J): I = 1,\cdots,d+1; J = 0, 1,\ldots,$

d+1}. Here I labels the current face of a triangulation cell
and J represents a random choice of the next face. However, for
the model "with exclusions" (the realistic case), successive J
values are constrained to be different, and we denote by $\bar{J}$ a
randomly chosen element of $\{0, \cdots, d+1\} \setminus \{J\}$.

$$(4.1) \qquad (I,J) \longrightarrow \begin{cases} \text{exit } + & \text{if} \quad I=1, \ J=0 \qquad = \text{Case} \quad \underline{1}^{\text{o}} \\ (I-1,\bar{0}) & \text{if} \quad I \neq 1, \ J=0 \qquad\qquad\quad \underline{2}^{\text{o}} \\ \text{exit } - & \text{if} \quad I=J=d+1 \qquad\qquad\ \underline{3}^{\text{o}} \\ (I+1,\bar{I}) & \text{if} \quad I=J \neq d+1 \qquad\qquad \underline{4}^{\text{o}} \\ (I-1,\bar{J}) & \text{if} \quad J=I-1 \neq 0 \qquad\qquad \underline{5}^{\text{o}} \\ (I+1,\bar{J}) & \text{if} \quad I \neq J=d+1 \qquad\qquad \underline{6}^{\text{o}} \\ (I,\bar{J}) & \text{if} \quad J \neq I-1, I, 0, d+1 \qquad \underline{7}^{\text{o}} \end{cases}$$

Case $\underline{1}^{\text{o}}$ corresponds to a forward level-crossing in the homotopy
direction, Case $\underline{3}^{\text{o}}$ to a backward crossing. The face labelled
(d+1, d+1), orthogonal to the homotopy axis, is assumed to be the
initial face. Cases $\underline{2}^{\text{o}}$, $\underline{4}^{\text{o}}$, $\underline{5}^{\text{o}}$, $\underline{6}^{\text{o}}$, $\underline{7}^{\text{o}}$ correspond to exits from
the current cell either orthogonally or at a 45° angle to the homo-
topy direction.

　　Before proceeding to a detailed micro-analysis of triangulation
K_1 "without exclusions", we observe that the micro-analyses (with
or without exclusions) for triangulation J_1 turn out to be pro-
babilistically equivalent (under maximal randomness) to those for
K_1 taken two levels at a time. For this reason, the more complicated pivoting rules for J_1 need not be treated separately.

　　When there are no excluded pivots in the K_1 triangulation,
the J index can be discarded from the state-space, and the type-I
transitions (on the smaller state-space $S' \equiv \{I : I = 1, \cdots, d+1\}$)
become

$$1 \longrightarrow \begin{cases} \text{exit} + & \text{with probability} & 1/(d+2) \\ 2 & \text{with prob.} & 2/(d+2) \\ 1 & \text{with prob.} & (d-1)/(d+2) \end{cases}$$

$$(4.2) \quad d+1 \longrightarrow \begin{cases} \text{exit} - & \text{with prob.} & 1/(d+2) \\ d & \text{with prob.} & 2/(d+2) \\ d+1 & \text{with prob.} & (d-1)/(d+2) \end{cases}$$

$$I \longrightarrow \begin{cases} I+1 & \text{with prob.} & 2/(d+2) \\ I-1 & \text{with prob.} & 2/(d+2) \quad \text{for} \quad I \neq 1, \, d+1 \\ I & \text{with prob.} & (d-2)/(d+2) \end{cases}$$

For type-II transitions when there are no excluded micro-steps, the only change to be made in the description (4.2) is

$$(4.2') \qquad d+1 \quad \rightarrow \quad \begin{cases} d & \text{with prob.} & 2/(d+1) \\ d+1 & \text{with prob.} & (d-1)/(d+1) \end{cases}$$

We show how to calculate p, s, and t from type-I micro-analysis on (4.2). Let p_I denote the probability of exiting + before exiting -, starting from I, and with transitions given by (4.2). Let v_1 denote the corresponding expected number of transition steps to exit (+ or -); and let s_I, t_I respectively denote the conditional expected number of steps from I to exit given that the direction of first exit is +, -. Finally, let $q_I \equiv 1 - p_1$ and $r_I \equiv p_I s_I$, so that $t_I = (v_I - r_I)/q_I$. We are interested in solving for $p \equiv p_{d+1}$, $s \equiv s_{d+1}$, $t \equiv t_{d+1}$. Now the following difference-equation holds:

$$(4.3) \quad \begin{cases} p_1 = (d+2)^{-1}(1+2p_2+(d-1)p_1) \\ p_{d+1} = (d+2)^{-1}(2p_d+(d-1)p_{d+1}) \\ p_I = (d+2)^{-1}(2p_{I+1}+2p_{I-1}+(d-2)p_I), \quad I \neq 1, \, d+1 \end{cases}$$

To understand (4.3), note for example from (4.2) that starting

from state 1, we exit + on the first step with probability
$1/(d+2)$; otherwise after the first step we start again from state
1 with probability $(d-1)/(d+2)$, retaining the same probability
P_1 of exiting + before -, or we go with probability $2/(d+2)$
to state 2 at the first step, after which the conditional probability of exiting + before - is P_2. Putting the three cases
together gives the first equation of (4.3).

    The system (4.3) can be re-written in matrix notation as

$$(4.4) \quad A\underline{p} = \begin{pmatrix} 1 \\ 0 \\ \cdot \\ \cdot \\ 0 \end{pmatrix}, \quad \underline{p} = \begin{pmatrix} P_1 \\ \cdot \\ \cdot \\ \cdot \\ P_{d+1} \end{pmatrix}, \quad A = \begin{pmatrix} 3 & -2 & & & & \\ -2 & 4 & -2 & & & \\ & -2 & 4 & -2 & & \\ & & \ddots & \ddots & \ddots & \\ & & & -2 & 4 & -2 \\ & & & & -2 & 3 \end{pmatrix}$$

Difference equations for $\underline{r} = \begin{pmatrix} r_1 \\ \cdot \\ \cdot \\ r_{d+1} \end{pmatrix}$ and $\underline{v} = \begin{pmatrix} v_1 \\ \cdot \\ \cdot \\ v_{d+1} \end{pmatrix}$ can be

similarly derived and re-written in the form

$$(4.5) \quad A\underline{v} = \begin{pmatrix} d+2 \\ \cdot \\ \cdot \\ d+2 \end{pmatrix}, \quad A\underline{r} = d\underline{p} \cdot (1 + 0(1/d)).$$

Now A can be inverted by solving a system of linear difference
equations, yielding

$$(4.6) \quad A^{-1} = \begin{pmatrix} \diagdown & \dfrac{(i+1)(d+3-j)}{2(d+4)}, \ j \geq i \\ & \diagdown \\ \dfrac{(j+1)(d+3-i)}{2(d+4)}, \ j < i & \diagdown \end{pmatrix}$$

$$(4.7) \quad p = p_{d+1} = \frac{2}{d+4}, \quad s = \frac{r_{d+1}}{p_{d+1}} \sim \frac{d^3}{12}, \quad v_{d+1} \sim \frac{d^2}{2},$$

$$t \sim \frac{v_{d+1} - r_{d+1}}{q_{d+1}} \sim \frac{d^2}{3}$$

Substituting the parameters from (4.7) into (2.5), (2.6), we find that the expectation and variance of total number of type-I steps required in the K_1 triangulation to reach $n+$ from $0+$ have leading terms (both in d and n)

$$(4.8) \qquad S_1(0) \sim n^2 d^3 / 4, \quad S_2(0) - S_1(0)^2 \sim n^4 d^6 / 24$$

Finally, the same micro-analysis for type-II steps using transitions (4.2') (still "without exclusions") leads to difference-equations exactly analogous to (4.5) but with matrix A replaced by

$$B = \begin{pmatrix} 3 & -2 & & & & \\ -2 & 4 & -2 & & & \\ & -2 & 4 & -2 & & \\ & & \ddots & \ddots & \ddots & \\ & & & -2 & 4 & -2 \\ & & & & -2 & 2 \end{pmatrix}, \quad B^{-1} = \begin{pmatrix} \diagdown & \dfrac{i+1}{2}, \ j \geq i \\ & \diagdown \\ \dfrac{j+1}{2}, \ j < i & \diagdown \end{pmatrix}$$

and it is easy to check that for type-II steps (using (2.7))

$$(4.9) \qquad \bar{s} \sim d^3 / 4, \quad S_1(0) \sim nd^3 / 4.$$

Although the analysis is much more complicated for transitions with excluded pivots as in (4.1), the asymptotic relations (4.7)–(4.9) remain valid.

5. DISCUSSION

The most evident conclusion of our model is the cubic growth (4.8) of convergence rate with dimension for the K_1 (and also other K) triangulations. It is easily calculated that the average directional density of Todd grows as $d^{3/2}$. These growth rates obviously differ, and both disagree with the little experimental evidence available. We informally consider the implications.

On a global level, average directional density is a measure of rate of convergence for curves of fixed length. That is, since the density measures number of pivots per length of the curve being followed, the $d^{3/2}$ growth can be expected only if the length of the curve is independent of dimension. Thus we can regard average directional density as a measure for the best case situation—the exponent 3/2 as a lower bound.

On the other hand, the cubic growth of (4.8) is a consequence of the assumption of maximal randomness—the worst case. Thus 3 is an upper bound on the exponent in the growth rate. Any actual problem will behave somewhere between the two extremes.

In particular, consider a carefully done experiment reported by R. Saigal at the Sardegna meeting. He considered a boundary-value problem which he formulated as a fixed-point problem in Banach space. Finite-dimensional approximations were made and numerically solved by piecewise linear homotopy methods. The convergence rate exhibited a growth d^e where $e \approx 1.75$. This particular problem is extremely well-behaved. The starting points of the homotopy lie in a region of convergence of Newton's method. That is, Newton's method converges rapidly to a solution, and thus so does the homotopy method. Hence the curve the homotopy method follows is very straight. The growth rate is, as expected, nearer the lower bound.

It would be interesting to conduct similar experiments on problems where the curve is more chaotic. L. Watson reports some of the curves in his applications are extremely wandering.

These observations indicate how our model should be improved. The assumption of maximal randomness is not generally valid. Each problem, or class of problems, has an associated local pivoting process that determines how wandering or non-wandering the path of convergence will be. Local structure could be obtained either by local theoretical analysis of the pivoting mechanism (e.g., by comparison with Newton's method) or by computational experience. The local transition probabilities will enter the final formula for convergence rate through the parameters p, s, t, etc., derived from micro-analysis. Further work along these lines is in progress.

REFERENCES

1. Feller, Introduction to Probability Theory and Applications, vol. I, 3rd ed. John Wiley (1968).

2. S. Karlin and H. Taylor, A First Course in Stochastic Processes, 2nd ed. Academic Press (1975).

3. G. van der Laan and A. J. J. Talman, Convergence and properties of recent variable dimension algorithms, in Numerical Solution of Highly Nonlinear Problems, W. Forster (ed.), North Holland Publishing Co., (1980), 3-36.

4. R. Saigal, Investigations into the efficiency of the fixed point algorithms, in Fixed Points, Algorithms and Applications, S. Karamardian, ed., Academic Press (1977), 203-223.

5. S. Smale, The fundamental theorem of algebra and complexity theory, Bull. A.M.S. 4(1981), 1-36.

6. M.J. Todd, The Computation of Fixed Points and Applications, Lecture Notes in Economics and Mathematical Systems #124, Springer (1976).

RELATIONSHIPS BETWEEN DEFLATION AND GLOBAL METHODS IN THE PROBLEM

OF APPROXIMATING ADDITIONAL ZEROS OF A SYSTEM OF NONLINEAR EQUATIONS

E.L. Allgower   K. Georg

Mathematics Dept.   Institut f. Angew. Mathematik
Colorado State University   Universität Bonn
Fort Collins Colo. 80523   5300 Bonn, West-Germany

ABSTRACT

In this paper we make a theoretical comparison of several methods which are used for numerically finding additional zeros of $F: \mathbb{R}^N \to \mathbb{R}^N$ (smooth with 0 a regular value) after one zero (say z^0) has already been obtained. In particular, we will compare the methods of deflation, global Newton, global homotopy and the d-trick.

§1. INTRODUCTION

The problem which we will discuss is the following: Given a zero-point z^0 of $F: \mathbb{R}^N \to \mathbb{R}^N$ (smooth with 0 a regular value), find additional zero-points of F. The problem of finding additional zero-points has applications in such diverse areas as nonconvex programming, nonlinear boundary and eigenvalue problems, electrical network problems, approximation theory and competitive equilibria.

One of the difficulties of this problem is that often z^0 has very strong "magnetic" properties relative to such iterative methods as Newton's method. This occurs when the convergence domain relative to the iterative method being used is very large in relationship to

This paper was written while the first author was a guest of the Deutsche Forschungsgemeinschaft at SFB 72, University of Bonn. Sponsored in part by AFOSR Grant FY 1456-81-00870.

the remaining zeros of F. Thus it is very difficult to obtain additional zero-points of F by the simple device of taking a new starting guess in the iterative process which is being used.

Before proceeding into our actual discussion, it might be noted
that there is a special case which now seems to be well in hand viz
the case of polynomial systems having low degree and low dimension.
This problem has recently been treated by several authors ([8],[10],
[13],[17]). The approach in these papers is to make a complex embedding via $x \rightarrow z \in C^N$ and then to construct an appropriate homotopy between F(z) and some chosen polynomial map P(z) whose zeros
are known. The numerical aspect then consists of tracing the homotopy paths which emanate from the known zeros of P(z). If however,
F involves transcendental functions or N is relatively large, the
number of extraneous complex zero-points which arise reduces the
practicality of this approach.

§2. DEFLATION

The deflation device has long been used in dimension N = 1 and
it has been frequently suggested and used by several writers (see
e.g. [11]). Brown and Gearhart [5] make a study of the performance
of deflation, but so far as the authors are aware, the performance
of deflation does not seem to be well understood from the theoretical
standpoint.

To try to find a second zero of F after z^0, one defines a
deflated mapping

(2.1) $f_{(1)}(x) = F(x) / \|x - z^0\|$

and applies an iterative zero finding method to $f_{(1)}$. If z^0 has
been found via an iterative method starting from x^0, then usually
x^0 is again used as a starting point and the same iterative method
is used. It is also clearly possible to repeatedly apply deflation
e.g.

(2.2) $f_{(k)}(x) = F(x) / \prod_{j=0}^{k-1} \|x - z^j\|$

where $z^0, z^1, \ldots, z^{k-1}$ are the successive zeros of F which have
been found using the same iterative process and starting point x^0.

In their study of deflation, Brown and Gearhart performed deflations with the ℓ_2, ℓ_∞ and a "gradient deflation"

$$f_{(k)}(x) = F(x) / \prod_{j=0}^{k-1} <DF(z_j), x - z^j> .$$

The iterative methods which were used were Newton's method and Brown's method [4]. The numerical examples given by Brown and Gearhart [5] showed that the performance of the deflation device varied very strongly according as the norm and iterative process were changed. This manifestation suggests that although z^o no longer has a domain of convergence, it is not in general necessary that x^o now belongs to the domain of convergence of some zero-point of $f_{(1)}$.

Since we are primarily concerned here with contrasting deflation and global methods, we will confine our interest concerning deflation to the combination of ℓ_2-norm and Newton's method. At the conclusion of our discussion of global methods we shall give several numerical examples.

§3. GLOBAL NEWTON METHODS

The global Newton methods have been developed and investigated primarily by Branin [2],[3] and by Smale [14],[19]. Applications to electrical network problems have been made by Chua and Ushida [9] and by Chao et al. [6],[7].

Global Newton methods involve tracing an integral curve of a differential equation such as

(GN) $\qquad \dot{x} = (\text{sign det } DF(x))(DF(x))^{-1}F(x)$

$\qquad x(0) = x^o ,$

where $\dot{x}$ may be regarded as the derivative of x with respect to arc length and $F: \mathbb{R}^N \to \mathbb{R}^N$ is the smooth map whose zeros are being sought. The integral curve for (GN) is in general, a smooth path in $\mathbb{R}^N$:

$$P(\text{GN}) = \{x(s) \in \mathbb{R}^N \mid (\text{GN}) \text{ is satisfied for } s \in [0,\infty)\} .$$

The classical Newton's method can be contrasted to the global Newton's method (GN) by interpreting it as the numerical integration of the classical Newton equation

$$\dot{x} = (DF(x))^{-1}F(x) \quad , \quad x(0) = x^{o}$$

via Euler's method using unit steps with respect to the arc length parameter. Hence it is generally the case that when x^{o} lies in the convergence domain of z^{o} for the classical Newton's method, then z^{o} will be the first zero-point of F which will be encountered along $p(GN)$.

A sufficient condition for the convergence of (GN) is given by the following

Boundary Condition   (Smale [19])

If $C \subset \mathbb{R}^{N}$ is a bounded connected domain having ∂C connected and smooth, and if

(BC) (i)   for all $x \in \partial C$, $DF(x)^{-1}$ exists and $DF(x)^{-1}F(x)$ always either points into C or always points out of C

      (ii) x^{o} is a regular starting point in ∂C i.e. $p(GN)$ is defined and smooth for $s \in [0,\infty)$,

then $p(GN) \cap \bar{C}$ contains an odd number of zeros of F.

By our assumption that F was smooth, that 0 was a regular value of F, and C is bounded, a result of Percell [18] allows the conclusion that (BC)(ii) is satisfied for almost all $x^{o} \in \partial C$. The condition (BC)(i) can be verified for any sufficiently small ball C about any zero-point of F.

Since our present aim is to locate as many zeros of F as possible, it may be reasonable in the search for additional zeros, to also numerically trace the integral curve defined by (GN) outward from C until it seems apparent that no additional zeros of F will be encountered.

§4. GLOBAL HOMOTOPY METHODS

The global homotopy methods have been developed and studied by Garcia and Gould [12] , Garcia and Zangwill [13] and by Keller [15], [16]. The idea here is to define a global homotopy such as

(GH) $\qquad H(t,x) = F(x) - tF(x^o) \qquad (x^o \in \text{dom } F)$

and to trace the homotopy path $P(GH)$ which is the connected component of $H^{-1}(0)$ containing $(1,x^o)$. Then any point $(0,z) \in P(GH)$
also satisfies $F(z) = 0$.

Let us assume that x^o has been chosen so that 0 is a regular value of H. Then by the implicit function theorem, $P(GH)$ is
a smooth curve and

(R) $\qquad \text{rank}(-F(x^o),DF(x)) = N \qquad \text{on} \quad P(GH)$.

It turns out that the global Newton and global homotopy methods are
very closely related, and in fact the paths $P(GN)$ and $P(GH)$ contain the same zeros of F (Garcia and Gould [12], Keller [16]).
An easy way to see this is to perform a differentiation with respect
to s of the equation $H(t,x) = 0$, which implicitly defines $P(GH)$.
Then we obtain the Davidenko initial value problem

(DIVP) $\qquad \begin{pmatrix} \dot{t} & \dot{x}^T \\ -F(x^o) & DF(x) \end{pmatrix} \begin{pmatrix} \dot{t} \\ \dot{x} \end{pmatrix} = \begin{pmatrix} 1 \\ 0 \end{pmatrix} , \quad \begin{pmatrix} t(0) \\ x(0) \end{pmatrix} = \begin{pmatrix} 1 \\ x^o \end{pmatrix} .$

By (R), and since $-F(x^o) = F(x)/t$ on $P(GH)$, we have that

$$\det \begin{pmatrix} \dot{t} & \dot{x}^T \\ \dfrac{F(x)}{t} & DF(x) \end{pmatrix} \neq 0 \qquad \text{on} \quad P(GH) .$$

Since

$$\begin{pmatrix} \dot{t} & \dot{x}^T \\ \dfrac{F(x)}{t} & DF(x) \end{pmatrix} \begin{pmatrix} \dot{t} & 0^T \\ \dot{x} & I \end{pmatrix} = \begin{pmatrix} 1 & * \\ 0 & DF(x) \end{pmatrix} ,$$

we have that $\det DF(x)$ changes sign exactly when $\dot{t}$ changes sign.
Thus if $(t(s),x(s))$ lies on $P(GH)$, then

$\qquad DF(x(s))\dot{x} - \dot{t}F(x^o) = 0 \qquad\qquad \text{hence}$

$\qquad DF(x(s))\dot{x} = \pm \text{ sign det } DF(x(s))F(x(s))/t(s)$

$\qquad\qquad t(s)\dot{x} = \pm \text{ sign det } DF(x(s)) (DF(x(s)))^{-1}F(x(s))$.

Hence $x(s) \in P(GN)$. Similarly, $x(s) \in P(GH)$ implies $(t(s),x(s)) \in P(GH)$.

The boundary condition for (GN) also has an analogue for (GH) ([12],[16]):

Let F, x°, C be as in (BC)(i)(ii). Then there exists an $L > 1$ such that

$$P(GH) \cap ([0,L] \times \partial C \cup \{L\} \times C) = \{(1,x^{\circ})\} .$$

From this it follows immediately that $P(GH)$ must assume the value $t = 0$ an odd number of times in $\bar{C}$.

§5. D-TRICK

In the d-trick [1], the homotopy

(H_d) $H_d(t,x) = F(x) - td$ with $\|d\| \neq 0$

is used to implicitly define a path $P(H_d)$ which contains the point $(0,z^{\circ})$ where $F(z^{\circ}) = 0$. This device will yield at least one more zero-point of F if a coercivity condition holds:

(CC)   There exists a bounded open neighborhood V of z° such that to each $x \in \partial V$, there corresponds a $\psi_x \in \mathbb{R}^N$ satisfying $\psi_x^T F(x) > 0$ and $\psi_x^T d < 0$.

Then obviously $P(H_d) \cap [0,\infty) \times \partial V = \emptyset$. Furthermore, on $\bar{V}$,

$$t = F_i(x)/d_i \qquad \text{for } d_i \neq 0 .$$

Hence $P(H_d)$ must be bounded in the positive t-direction. By the parametrized Sard's theorem (see e.g. [8]) $P(H_d)$ is a smooth curve for almost all choices for $d \in \mathbb{R}^N$. Since z° is a regular point of F, $(0,z^{\circ})$ is a regular point of H_d. Hence $P(H_d)$ must inter-sect $\{0\} \times V$ at some point $(0,z^1)$ with $z^1 \neq z^{\circ}$.

The d-trick can be regarded as a generalization of the idea of following $P(GH)$ backward from $(0,z^{\circ})$ i.e. outward from ∂C at

$(1,x^o)$. The main difference is that $d \in \mathbb{R}^N$ need not equal $F(x^o)$, or indeed, d need not even belong to range F .

In [1] several concrete examples are given in which a coercivity condition can easily be verified. As an example, consider an F such that some component F_i satisfies $F_i(x) > 0$ for $\|x - z^o\| = r$, then

$$d = -e_i \qquad \text{with } \Psi_x \equiv e_i$$

will be adequate for coercivity on $\|x - z^o\| = r$.

Often it is possible to make a choice of d such that $P(H_d)$ is in fact a closed loop. The global homotopy and d-trick can be combined. If following $P(GH)$ backward from $(0,z^o)$ doesn't lead to a new zero-point, $P(H_d)$ might do so. Finally, let us remark that successive zero-points which are found along $P(GH)$, $P(GN)$ or $P(H_d)$ have opposite index.

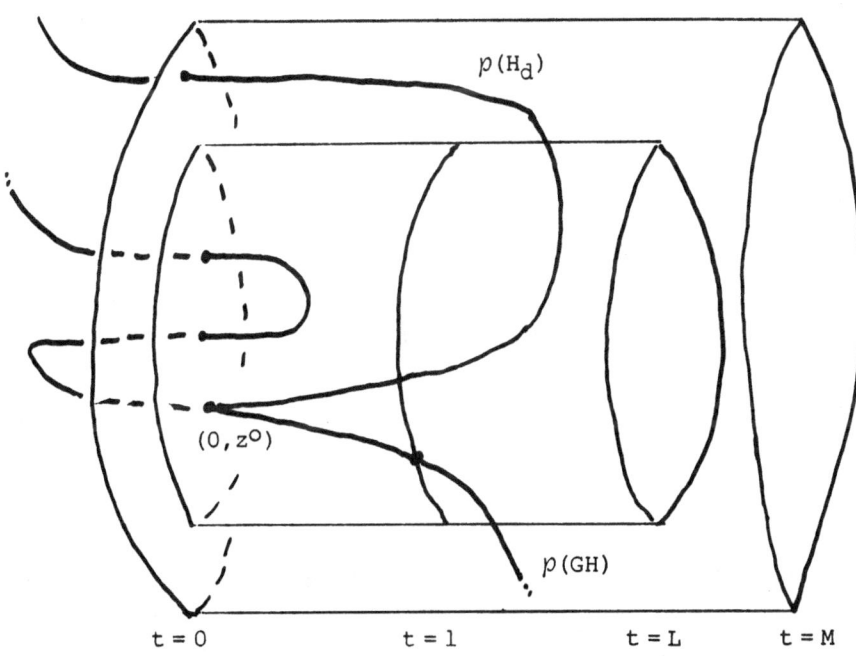

§6. GLOBAL DEFLATION

It might seem tempting to consider a combination of deflation and global homotopy in order to seek additional zero-points of F. Thus we could consider a global deflation homotopy

$$(GD) \qquad G(t,x) = \frac{F(x)}{\|x - z^{\circ}\|} - t \frac{F(x^{\circ})}{\|x^{\circ} - z^{\circ}\|} \,.$$

Then we may consider the homotopy path $P(GD) \subset G^{-1}(0)$ which contains $(1,x^{\circ})$.

So long as $x \neq z^{\circ}$, we could just as well consider the modified global deflation homotopy

$$(\widetilde{GD}) \qquad \widetilde{G}(t,x) = F(x) - t \frac{\|x - z^{\circ}\|}{\|x^{\circ} - z^{\circ}\|} F(x^{\circ})$$

and the corresponding connected set $P(\widetilde{GD}) \subset \widetilde{G}^{-1}(0)$ which contains $(1,x^{\circ})$.

Technically $P(\widetilde{GD})$ is not a smooth path, but the union of $P(GD)$ and $\mathbb{R} \times \{z^{\circ}\}$. However, this is unimportant since we are only interested in $P(\widetilde{GD})$ away from $x = z^{\circ}$.

By comparing $P(GH)$ and $P(\widetilde{GD})$ however, we immediately see that if

$$\left(t(s) \frac{\|x(s) - z^{\circ}\|}{\|x^{\circ} - z^{\circ}\|} , x(s) \right) \in P(\widetilde{GD}) ,$$

then also

$$(t(s),x(s)) \in P(GH) .$$

Thus in particular, the homotopy paths $P(GD)$ or $P(\widetilde{GD})$ can yield no other zeros of F in addition to those which can be found by tracing $P(GH)$ in both directions starting from $(1,x^{\circ})$. The same remark applies of course if $P(GN)$ is traced in both directions from x°.

Since (GD) represents a "globalization" of the standard deflation, we can expect that $P(GD)$ will yield the zeros which standard deflation might reach. In fact, comparison with the deflation examples of Brown and Gearhart [5] shows that $P(GH)$ and $P(H_d)$ often succeeds in reaching more zeros of F than standard deflation does. It is however, conceivable that for some higher order deflation $f_k(x)$, the starting point x^O might accidently yield convergence to a zero-point of F which is not accessible via $P(GH)$. In fact however, the general experience seems to be that deflation diverges before all of the zeros of F on $P(GH)$ are reached. As a further confirmation of the relationship of deflation to the global methods, we note that in the numerical examples of [5] (using ℓ_2- norm and Newton's method) when successive zero-points of F are found, they have opposite index.

§7. EXAMPLES [5] (ℓ_2-norm and Newton's method)

I. The cubic-parabola

$$
F(x,y) = \begin{pmatrix} 4x^3 - 3x - y \\ x^2 - y \end{pmatrix}
$$

The zeros of F are $z^O = (1,1)^T$, $z^1 = (0,0)^T$, and $z^2 = (-3/4 , 9/16)^T$. It is reported that z^O is a highly magnetic zero, and that deflation using z^O yielded either z^1 or divergence. A routine calculation shows $\text{ind}(F,z^O) = -1 = \text{ind}(F,z^2)$, $\text{ind}(F,z^1) = 1$. Hence it is impossible for a homotopy path to reach z^2 directly from z^O without first reaching z^1. By using the d-trick with $d = (1,0)^T$, we can easily see that $P(H_d)$ passes through all three zero-points.

II. The four-cluster

$$
F(x,y) = \begin{pmatrix} (x - y^2)(x - \sin y) \\ (\cos y - x)(y - \cos x) \end{pmatrix}
$$

There are 4 zeros which are nearly equal:
$z^O \doteq (.67,.82)^T, z^1 \doteq (.64,.80)^T , z^2 \doteq (.70,.78)^T , z^3 \doteq (.69,.77)^T$
plus others which are farther away. It is reported that with

$x^O = (.9,1)^T$, z^O and then z^1 were found and in no case were more
than 2 zero-points from the four-cluster found. Other zeros elsewhere however, were sometimes found.

Since

$$\text{ind}(F,z^O) = 1 \quad \text{and} \quad \text{ind}(F,z^1) = \text{ind}(F,z^2) = \text{ind}(F,z^3) = -1 ,$$

it is utterly impossible for a homotopy path to obtain more than 2
points of the four-cluster without having first reached some other
zero-point of F .

III. The hyperbola-circle

$$F(x,y) = \begin{pmatrix} xy - 1 \\ x^2 + y^2 - 4 \end{pmatrix} , \quad x^O = (0,1)^T .$$

The zero-points of F are $\pm(2 + \sqrt{3} , 1/(2 + \sqrt{3}))^T$ and
$\pm(2 - \sqrt{3} , 1/(2 - \sqrt{3}))^T$. It is reported that only the two positive
zeros were found from x^O.

It is routine to see that the global degree of F is zero and
that $P(GH)$ with $x^O = (0,1)^T$ is a closed loop containing all four
zero-points.

IV. The 3×3 system

$$F(x,y,z) = \begin{pmatrix} x^2 + 2y^2 - 4 \\ x^2 + y^2 + z - 8 \\ (x - 1)^2 + (2y - \sqrt{2})^2 + (z - 5)^2 - 4 \end{pmatrix}$$

has zero-points $z^O = (0,\sqrt{2},6)^T$, $z^1 = (2,0,4)^T$. It is reported
that with $x^O = (1,1,1)^T$, z^O but not z^1 was found.

It is routine to see that if the global homotopy path $P(GH)$
for

$$H(t,x,y,z) = F(x,y,z) - tF(1,1,1)$$

is followed beyond $(0,\sqrt{2},6)^T$, then it will eventually turn back
upward and reach $(2,0,4)^T$.

BIBLIOGRAPHY

[1]   Allgower, E. L. and Georg, K., Homotopy methods for approxi-
         mating several solutions to nonlinear systems of equations,
         in: Numerical Solution of Highly Nonlinear Problems,
         ed. W. Forster, North-Holland, Amsterdam, 1980, 253-270.
[2]   Branin, F. H. Jr., Widely convergent method for finding multiple
         solutions of simultaneous nonlinear equations, I.B.M. J.
         Research Develop. 16 (1972), 504-522.
[3]   Branin, F. H. Jr. and Hoo, S. K., A method for finding multiple
         extrema of a function of N variables, Proceedings of the
         Conference on Numerical Methods for Nonlinear Optimisation,
         University of Dundee, Scotland, 1971; Numerical Methods
         for Nonlinear Optimisation, ed. F. A. Lootsma, Academic
         Press, London, 1972, 231-237.
[4]   Brown, K. M., A quadratically convergent Newton-like method
         based upon Gaussian elimination, SIAM J. Numer. Anal. 6
         (1969), 560-569.
[5]   Brown, K. M. and Gearhart, W. B., Deflation techniques for the
         calculation of further solutions of a nonlinear system,
         Numer. Math. 16 (1971), 334-342.
[6]   Chao, K. S., Liu, D. K. and Pan, C. T., A systematic search
         method for obtaining multiple solutions of simultaneous
         nonlinear equations, IEEE Transactions on Circuits and
         Systems, CAS-22 (1975), 748-753.
[7]   Chao, K. S. and Saeks, R., Continuation methods in circuit
         analysis, Proc. of the IEEE 65 (1977), 1187-1194.
[8]   Chow, S. N., Mallet-Paret, J. and Yorke, J. A., A homotopy
         method for locating all zeros of a system of polynomials,
         in: Functional Differential Equations and Approximation of
         Fixed Points, eds. H. O. Peitgen and H. O. Walther, Springer
         Lecture Notes in Math. 730 (1979), 77-88.
[9]   Chua, L. O. and Ushida, A., A switching-parameter algorithm
         for finding multiple solutions of nonlinear resistive cir-
         cuits, IEEE Trans. Circuit Theory and Applications 4 (1976),
         215-239.
[10] Drexler, F. J., Eine Methode zur Berechnung sämtlicher Lösungen
         von Polynomgleichungssystemen, Numer. Math. 29 (1977),
         45-58.
[11] Forsythe, G. and Moler, C. B., Computer solution of linear
         algebraic systems, Prentice-Hall, Englewood Cliffs N. J.,
         1967.
[12] Garcia, C. B. and Gould, F. J., Relations between several path
         following algorithms and local and global Newton methods,
         SIAM Review 22 (1980), 263-274.
[13] Garcia, C. B. and Zangwill, W. I., Finding all solutions to
         polynomial systems and other systems of equations, Math.
         Programming 16 (1979), 159-176.
[14] Hirsch, M. and Smale, S., On algorithms for solving f(x) = 0.,
         Comm. Pure Appl. Math. 32 (1979), 281-312.

[15]   Keller, H. B., Numerical solution of bifurcation and nonlinear
       eigenvalue problems, in: Applications of Bifurcation Theory,
       ed. P. Rabinowitz, Academic Press, New York, 1977, 359-384.

[16]   Keller, H. B., Global homotopies and Newton methods, in:
       Recent Advances in Numerical Analysis, eds. C. de Boor and
       G. H. Golub, Academic Press, New York, 1978, 73-94.

[17]   Kojima, M., Nishino, H. and Arima, N., A PL homotopy for
       finding all the roots of a polynomial, Math. Programming
       16 (1979), 37-62.

[18]   Percell, P., Note on a global homotopy, Numer. Funct. Anal.
       and Optim. 2 (1980), 99-106.

[19]   Smale, S., A convergent process of price adjustment and global
       Newton methods, J. Math. Econ. 3 (1976), 107-120.

SMOOTH HOMOTOPIES FOR FINDING ZEROS

OF ENTIRE FUNCTIONS

Jack Carr and John Mallet-Paret

Heriot-Watt University, Edinburgh, Scotland and

Michigan State University, East Lansing, Michigan

§1. Introduction

We study smooth families (homotopies) of entire functions
$f : C \times [0,1] \to C$:

$f_t(z) = f(z,t)$, smooth for $(z,t) \in C \times [0,1]$;

f_t is entire (i.e. analytic for all $z \in C$, the
complex plane) for each t.

Given an entire function f_1, we wish to construct such a
homotopy satisfying the properties that

(1) f_0 is trivial enough that all its zeros a_k
are known;

(2) from each a_k there is a path $\alpha_k(t)$ of zeros
of f_t (so $f_t(\alpha_k(t)) = 0$ and $\alpha_k(0) = a_k$)
extending throughout $0 \le t \le 1$, where $\alpha_k(1) = b_n$
is some zero of f_1; and

43

(3)   for each zero b_n of f_1, there exists some

path $\alpha_k(t)$ reaching it, as in (2).

By (numerically) following these paths, we in some sense locate

"all" zeros of f_1. In [5] homotopies were described which

would locate some zero of a smooth map in R^n. This has been

numerically implemented by Watson, who solves an associated

differential equation as described in [5]. In [6] and [8]

systems of polynomials in C^n were considered; here, all zeros

were found, as above (not the case in R^n); moreover, as above,

the paths $\alpha(t)$ were monotone in t. Our present work is in

some sense an extension of [6] to entire, or transcendental

functions. However, here we are motivated by classes of functions

arising in a specific setting, namely calculation of eigenvalues

of linear boundary value problems. Locating such values is of

central importance in many problems of engineering, and of other

areas, where stability and bifurcation of both linear and nonlinear

systems are studied. Boundary value problems in mechanics (such

as Beck's problem [2], [3], [4], [10]) as well as reaction –

diffusion systems of chemical engineering (see [7]) are typical

problems. In Section 3 we mention several examples in more detail.

In what follows we shall only indicate the main ideas of the

proofs of our results and examples; details will be presented

later.

§2. Homotopies of Entire Functions

A homotopy $f(z,t)$ is <u>regular</u> if 0 is a regular value of f for $t < 1$ (where f is regarded as a mapping from $(z,t) = (x + iy, t) \in R^3$ to $f = g + ih \in R^2$), and if also f_1 is not identically zero. Thus 0 is such a regular value if $f(z,t) = 0$ and $t < 1$ implies the matrix $\frac{\partial(g,h)}{\partial(x,y,t)}$ has maximal rank two. By the Cauchy–Riemann equations $(\frac{\partial g}{\partial x} = \frac{\partial h}{\partial y}$ and $\frac{\partial g}{\partial y} = -\frac{\partial h}{\partial x})$, $\frac{\partial(g,h)}{\partial(x,y,t)}$ has rank two if and only if $\frac{\partial(g,h)}{\partial(x,y)}$ is not zero; or equivalently, the complex number $\frac{\partial f}{\partial z}(z,t) \neq 0$. Thus 0 is a regular value of f for $t < 1$ if and only if

$$f(z,t) = 0 \quad \text{and} \quad t < 1 \quad \text{implies} \quad \frac{\partial f}{\partial z}(z,t) \neq 0 \; .$$

Regular homotopies have the property that their zeros lie on smooth curves $z = \alpha(t)$ which either tend to infinity or join up with some zero $(z,t) = (a_k, 0)$ of f_0 as t decreases; similar behavior holds for increasing t where the curves may join up with zeros of f_1. Indeed, the arguments of [6], based mainly on the implicit function theorem, and analyticity in z, easily show the following facts for regular homotopies.

(1) Through any zero (z_0, t_0), $t_0 < 1$, of f, there exists a unique smooth curve $z = \alpha(t)$ of zeros, (that is, $f(\alpha(t), t) = 0$, $\alpha(t_0) = z_0)$ at least for t near t_0. In particular this is so at each zero $(a_k, 0)$; we let $\alpha_k(t)$ denote the curve of zeros through this point.

(2) Any two curves $\alpha(t)$ and $\beta(t)$ of zeros of f

on overlapping intervals I_α and I_β which

agree at one point of $I_\alpha \cap I_\beta$ must agree

everywhere there. This implies any curve of

zeros has a unique extension to a maximal

interval in $[0,1]$.

(3) If $\sigma \in [0,1)$ is the left-hand endpoint of the

maximal interval for the curve $\alpha(t)$, then either

$|\alpha(t)| \to \infty$ as $t \downarrow \sigma$, or $\alpha(t) \to a_k$ for some

k, in which case $\sigma = 0$ and $\alpha(t) = \alpha_k(t)$. Near

a right hand endpoint τ, either $|\alpha(t)| \to \infty$ or,

for some n, $\alpha(t) \to b_n$ as $t \uparrow \tau = 1$.

An argument using Rouche's theorem also shows

(4) if b_n is a zero of f_1, then there exists

a curve of zeros $\beta_n(t)$, for t near 1, with

$\beta_n(t) \to b_n$ as $t \uparrow 1$. (Such a curve may not

be unique; there are as many such curves as the

order of the zero b_n of f_1.)

We say $f(z,t)$ is a <u>regular equivalence</u> if it is regular,

if each path $\alpha_k(t)$ from a_k reaches some b_n as $t \uparrow 1$, and

if each b_n is such a limit of some $\alpha_k(t)$. Constructing

homotopies which are regular generally involves a straightforward

application of Sard's theorem, as in [5] and [6]. We do not

address this problem here. The main problem we do consider is determining when a regular homotopy is a regular equivalence. In

other words, when can the behavior of zeros $\alpha_k(t) \to \infty$ or

$\beta_n(t) \to \infty$ be excluded? If this is done, then differentiating

$f(\alpha(t),t) = 0$ shows each path $z = \alpha_k(t)$ may be calculated as

the solution of the initial value problem

$$\frac{\partial z}{\partial t} = -\frac{\partial f}{\partial t} \Big/ \frac{\partial f}{\partial z} \,,$$

$$z(0) = a_k \,,$$

and each zero b_n is obtained as the value $z(1)$ of some

solution. Below we define a class of functions $\mathcal{F}$ and

homotopies $\mathcal{K}$ within $\mathcal{F}$ for which every regular homotopy is a

regular equivalence.

Consider entire functions of the form

$$(2.1) \qquad f(z) = \sum_{k=1}^{N} h_k(z^{-1}) z^{P_k} e^{c_k z} \qquad \text{for } |z| \text{ large}$$

where $h_k(w)$ is analytic in a neighborhood of $w = 0$, P_k is

an integer (not necessarily positive), and c_k is complex. If

we assume (which we may without loss) that

$$h_k(0) \neq 0 \quad \text{for all } k, \quad \text{and}$$

$$c_j \neq c_k \quad \text{if } j \neq k \,,$$

we say f is in normal form, $\{c_k\}$ are the nodes of f, and

P_k is the weight of c_k. Also, for f in normal form define the nodal polygon $\Pi(f) \subseteq C$ as

$$\Pi(f) = \text{convex hull of the complex}$$

$$\text{conjugates } \bar{c}_k \text{ of the nodes .}$$

We distinguish vertex nodes, edge nodes, and interior nodes

depending on whether $\bar{c}_k$ lies on a vertex, an edge (but not a

vertex), or inside of $\Pi(f)$. (If $\Pi(f)$ is an interval $[\bar{c}_1, \bar{c}_2]$,

then c_1 and c_2 are vertex nodes, and any other c_k are edge

nodes.) The class $\mathcal{F}$ is now defined as those entire functions

as above satisfying also:

   (1) $N \geq 2$, so there is more than one node;

   (2) all vertex nodes have the same weight p (called

        the vertex weight); and

   (3) $p_k \leq p$ for every weight p_k of an edge node.

Functions in $\mathcal{F}$ enjoy the property that their zeros have a

rather nice asymptotic distribution as they tend to infinity:

they all lie along "root chains", in semi-infinite strips

$$S(\theta,R) = \{z \in C \mid |z - se^{i\theta}| \leq R \text{ for some } s \geq 0\}$$

where θ is an asymptotic direction. The asymptotic directions

are determined from the nodal polygon: they are those $\theta \in [0,2\pi)$

such that the vector $e^{i\theta}$ is an outward normal to some edge of

$\Pi(f)$. The notion of root chain can be found in other sources, notably [1, Chap. 12] where the zero distribution of a somewhat

different class of exponential polynomials is studied. The

asymptotic distribution of eigenvalues of certain boundary value

problems is studied from this perspective in [9, Chap. XIX, Sec. 4].

We outline some of the ideas involved in proving these facts

about root chains; detailed proofs of this and other results

will appear separately. We point out here that the width R

of the strip can be explicitly estimated, and so provide a priori

estimates of the location of the roots. This should be useful

in implementation of the algorithm.

If $f(z) = 0$, we examine the magnitude of the terms

$T_k = h_k(z^{-1})z^{p_k}e^{c_k z}$ in the normal form (2.1). If T_{k_1} and

T_{k_2} are the terms of largest and second largest magnitude,

then the bound

(2.2) $$(N-1)^{-1} \leq \left| T_{k_2} T_{k_1}^{-1} \right| \leq 1$$

holds since (2.1) is a sum of N terms. Also $\left| T_k T_{k_1}^{-1} \right| \leq 1$ for

any k. For roots with $|z|$ large, (2.2) is of the order

$|z|^{p_{k_2}-p_{k_1}} \exp\{\text{Re}[(c_{k_2} - c_{k_1})e^{i\theta}]|z|\}$ where $z = |z|e^{i\theta}$.

Necessarily, $\text{Re}[(c_{k_2} - c_{k_1})e^{i\theta}] \to 0$ as $|z| \to \infty$ for this bound

to be maintained, implying the limiting vector $e^{i\theta}$ is orthogonal

to the segment $[\overline{c_{k_1}, c_{k_2}}]$. The one-sided bound on the ratio of

T_k and T_{k_1} shows further $Re[(c_k - c_{k_1})e^{i\theta}] \leq 0$ in the limit,

and this implies θ is an asymptotic direction. More refined

estimates involving the $|z|^{P_k}$ terms and conditions on the P_k

show in fact z lies in some strip $S(\theta, R)$.

Let us consider several examples of functions in $\mathcal{F}$.

Example. $\dfrac{\sin z}{z} = \dfrac{1}{2iz}(e^{iz} - e^{-iz})$ belongs to $\mathcal{F}$, with

$\Pi = [-i, i]$, and vertex weight $p = -1$.

Example. $z + \sin z$ does not belong to $\mathcal{F}$, as the nodes

$-i$, 0, i have weights 0, 1, 0 respectively.

Example. $z^{100} + \sin z \sinh 2z$ has vertex nodes $\pm 2 \pm i$

of weight 0, and an interior node 0 of weight 100, hence

belongs to $\mathcal{F}$.

Example. $\dfrac{\sin z}{z} + \sin cz$ belongs to $\mathcal{F}$ if $c > 1$ (with

vertex weight 0) but not if $0 < c < 1$.

Example. $f(z) = 2 \cosh [(z^2 - 1)^{1/2}]$ belongs to $\mathcal{F}$:

$\cosh t^{1/2} = \displaystyle\sum_{n=0}^{\infty} \dfrac{t^n}{(2n)!}$ is an entire function, hence so is f.

And for large $|z|$,

$$f(z) = h_1(z^{-1})e^z + h_2(z^{-1})e^{-z}$$

where

$$h_1(w) = \exp\{w^{-1}[(1-w^2)^{1/2} - 1]\}$$

$$h_2(w) = \exp\{-w^{-1}[(1-w^2)^{1/2} - 1]\}$$

are analytic and non-zero at $w = 0$. Observe that neither h_1

nor h_2 is analytic at $w = \pm 1$, even though f is.

Now we consider a class of homotopies $\mathcal{N}$ within $\mathcal{F}$. To

be precise, $f(z,t)$ is in $\mathcal{N}$ if

(1) $f_t \in \mathcal{F}$ for each $t \in [0,1]$;

(2)   near any $t_0 \in [0,1]$ we have for large $|z|$

$$f(z,t) = \sum_{k=1}^{N} h_k(z^{-1},t) z^{p_k} e^{c_k z}$$

(not necessarily in normal form), with

(i) $h_k(w,t)$ smooth in (w,t) and analytic in

w near $w = 0$, and with

(ii) p_k and c_k independent of t; and

(3)   the nodal polygon $\Pi(f_t)$ and vertex weight $p(f_t)$

are independent of $t \in [0,1]$.

We say "not necessarily in normal form" above, as the normal form

could change with t; for example $z + (z^{-1} + t) z e^z + z e^{2z}$ changes

at $t = 0$, yet this homotopy is in $\mathcal{N}$. Some other examples of

homotopies: $t z^{100} + \sin z \sinh 2z$ is in $\mathcal{N}$, but $t z^{100} + \sin z$

is not; and $t \sin c_1 z + \sin c_2 z$ is if $c_1 < c_2$, but not if

$c_1 > c_2$. Homotopies in $\mathcal{N}$ have the property that root chains exist uniformly, that is, independently of t. For this reason

we can rule out the possibility of a curve of zeros $z = \alpha(t)$

tending to infinity as t approaches some $\tau \in [0,1]$; if this

were to happen, then $\alpha(t) \to \infty$ along a particular strip $S(\theta,R)$;

and this would contradict the behavior of f at infinity along

$S(\theta,R)$. Suppose for example $\theta = 0$ were an asymptotic

direction; then along $S(0,R)$, simple estimates show that near

infinity $f(z) = z^p e^{\zeta z} (\Sigma' H_k e^{id_k z} + o(1))$ as $t \to \tau$, where

$H_k = h_k(0,\tau)$ and the sum Σ' involves only those terms for

which $p = p_k$, and $c_k = \zeta + id_k$ lies on the rightmost edge of

Π. If $\alpha(t) = \xi(t) + i\eta(t) \to \infty$ in $S(0,R)$, then $\xi(t) \to \infty$

while $\eta(t)$ is bounded; but arguments based on the analyticity

and almost periodicity of the sum Σ' show this cannot happen.

We have in fact the following result.

Proposition 1. Let f ϵ $\mathscr{N}$, and let $\alpha(t)$ be a continuous

path of zeros of f_t on some interval $(\sigma,\tau) \subseteq [0,1]$. Then

$\alpha(t)$ approaches finite limits as $t \downarrow \sigma$ and $t \uparrow \tau$.

An immediate consequence of this proposition is:

Theorem. Let f ϵ $\mathscr{N}$ be a regular homotopy. Then f is

a regular equivalence.

To illustrate these results, it is a simple exercise to

calculate the paths $\alpha(t)$ for the homotopy t sin z − sin 2z =

(sin z)(t − 2 cos z) in $\mathscr{N}$, and for

$\sin z - \frac{t}{4} \sin 2z = (\sin z)(1 - \frac{t}{2} \cos z)$ which is not in $\mathcal{N}$.   Both

are regular homotopies, so the first is a regular equivalence.

But the second has infinitely many curves $\alpha(t) = 2N\pi \pm i \text{ arccosh}(\frac{2}{t})$.

N = integer, tending to infinity as $t \downarrow 0$.   Because

$2N\pi \pm i \text{ arccosh } 2$ is a simple zero of $f(z,1)$, $\alpha(t)$ is the only

path leading to this point. Hence such zeros will not be obtained

by following paths from the zeros of $f(z,0) = \sin z$.

Another result classifies further the structure of the zeros

for $f \varepsilon \mathcal{N}$.

Proposition 2. Let $f \varepsilon \mathcal{N}$, and $f(z_0, t_0) = 0$ for some

$(z_0, t_0) \varepsilon C \times [0,1]$. Then there exists a continuous path

$z = \alpha(t)$, $0 \le t \le 1$, of zeros of f_t satisfying $\alpha(t_0) = z_0$.

Here, continuous path means just that - $\alpha(t)$ is a

continuous, but not necessarily smooth function of t. The zero

set of f could be very complicated, involving paths with

branches and bifurcations. But because of the analyticity in z,

a continuous path can be singled out.

§3. Examples and Applications

We consider here boundary value problems for systems of

linear constant coefficient ordinary differential equations. The

coefficients and boundary conditions depend on a complex

parameter λ, and one seeks those values of λ for which the problem has a non-zero solution. Generally, such problems

arise from linear partial differential equations upon separation

of variables, where λ is an eigenvalue related to the stability

of the PDE, (Re $\lambda < 0$ for all solutions λ being a typical

stability condition.)

Often, additional parameters $t_1, t_2, \ldots$ are present in the

problem, so the solutions λ may be thought of as depending on

these t_k. An important problem is to determine values of t_k

at which Re $\lambda = 0$ for some λ, as this can indicate a change

in stability. It is natural therefore to regard the t_k as

homotopy parameters, and to follow λ as they vary.

Below we present several examples of such problems. The

general approach is to solve the ODE explicitly and impose the

associated boundary conditions; this reduces the problem to

finding the zeros of some known function of λ. For a wide

variety of important problems this function is either in $\mathcal{F}$, or

is equivalent to a function in $\mathcal{F}$ via a simple change of

variables. Finally, a homotopy in $\mathcal{N}$ to a trivial function can

be constructed. Alternatively, the original parameters t_k of

the problem may be used as homotopy parameters; in this case

we must determine whether or not such a homotopy is in $\mathcal{N}$.

Consider the boundary value problem

$$y''(x) = \lambda y(x)$$
(3.1)
$$y(0) = 0, \quad y'(1) = ty(1)$$

where we seek the eigenvalues λ, for a fixed parameter

$t \in [0,1]$. For $\lambda \neq 0$ the general solution of $y'' = \lambda y$ is

$y = A \sinh(\lambda^{1/2}x) + B \cosh(\lambda^{1/2}x)$, for constants A and B.

There is a nontrivial solution $(A \neq 0$ or $B \neq 0)$ of the

boundary value problem (3.1) if and only if $\cosh \lambda^{1/2} =$

$t\lambda^{-1/2}\sinh \lambda^{1/2}$. Letting $z = \lambda^{1/2}$ we may write this as

$f(z,t) = 0$ where

$$f(z,t) = \cosh z - tz^{-1}\sinh z$$

$$= \frac{1}{2}(1 - tz^{-1})e^{z} + \frac{1}{2}(1 + tz^{-1})e^{-z}$$

Certainly $f \in \mathcal{N}$, with $\Pi = [-1,1]$ and vertex weight $p = 0$.

The zeros of f_0 are $a_k = (k+\frac{1}{2})\pi i$ where k is an integer;

all these zeros are simple. By letting $z = iy$, y real, we

see the curves of zeros $z = \alpha_k(t)$ explicitly as the inter-

section points of the graph of $\tan y = \dfrac{\sinh iy}{i \cosh iy}$ with the

line of slope t^{-1} (for $t > 0$; of course as $t \to 0$, $y \to$ some

a_k). These are curves of simple zeros, and extend for $t \in [0,1]$.

They account for all the zeros, since by Proposition 2, any other

zero (z_0,t_0) would be connected to some a_k at $t = 0$; but

the curve $\alpha_k(t)$ extending from a_k is unique as it is simple.

Thus f is a regular homotopy, hence a regular equivalence.

Now consider the problem

$$y''(x) = \lambda y(x)$$

$$y(0) = 0, \quad ty'(1) = y(1) \ .$$

As above this leads to the homotopy

$$f(z,t) = t \cosh z - z^{-1} \sinh z$$

$$= \frac{1}{2}(t - z^{-1})e^z + \frac{1}{2}(t + z^{-1})e^{-z} \ .$$

Here $f \notin \mathcal{N}$; although $\Pi = [-1,1]$ throughout the homotopy, the vertex weight drops from $p = 0$ for $t > 0$ to $p = -1$ at $t = 0$. In fact, f has curves of zeros $z = \pm\beta(t)$, $0 < t \le 1$, with $|\beta(t)| \to \infty$ as $t \to 0$. This is seen by considering z real, and intersecting the graph of $\tanh z$ with the line of slope t; as $t \downarrow 0$, the two intersection points move horizontally to infinity. The roots $\pm\beta(1)$ therefore would never be obtained from a homotopy beginning at $t = 0$.

Beck's Problem is described by the partial differential equation

$$u_{ss} + u_{xxxx} + qu_{xx} = 0 \ ,$$

$$u(s,0) = u_x(s,0) = u_{xx}(s,1) = u_{xxx}(s,1) = 0 \ ,$$

for $u(s,x)$, where s is time, $x \in [0,1]$, and $q \ge 0$ is a constant. This represents the displacement of a thin elastic rod, fixed at $x = 0$, and subjected at $x = 1$ to a compressive tangential load of magnitude q. To determine stability of this problem, set $u = e^{\lambda s}y(x)$ to obtain

(3.2) $$y''' + qy'' + \lambda^2 y = 0$$

(3.3) $$y(0) = y'(0) = y''(1) = y'''(1) = 0 .$$

We shall obtain a homotopy $f(z,q)$ in $\mathscr{X}$ on any interval

$[0,q_0]$, where $\lambda = z^2$. A condition for stability is $\mathrm{Re}\ \lambda \le 0$

for every zero λ, with possibly finitely many exceptions,

which can be checked separately.

To obtain f, consider the polynomial

(3.4) $$m^4 + qm^2 + z^4 = 0$$

which has distinct roots m_1, $-m_1$, m_2, and $-m_2$ if and only if

(3.5) $$z^4 \ne 0, \quad \frac{q^2}{4} ;$$

in this case, the general solution of the differential equation

(3.2) is

$$y(x) = A_1 \sinh m_1 x + A_2 \sinh m_2 x + B_1 \cosh m_1 x$$

$$+ B_2 \cosh m_2 x .$$

Substitution into (3.3) shows these boundary conditions are

satisfied for a non-trivial solution if and only if

$$m_1^4 + m_2^4 - 2m_1^2 m_2^2 \cosh m_1 \cosh m_2$$

$$+ (m_1^2 + m_2^2)m_1 m_2 \sinh m_1 \sinh m_2 = 0 .$$

Define now

$$\phi(\mu_1,\mu_2) = \mu_1^4 + \mu_2^4 - 2\mu_1^2\mu_2^2 \cosh \mu_1 \cosh \mu_2$$

(3.6)
$$+ (\mu_1^2 + \mu_2^2)\mu_1\mu_2 \sinh \mu_1 \sinh \mu_2,$$

$$f(z,q) = \phi(m_1,m_2)$$

and note the following:

   (1) ϕ is analytic for all μ_1, μ_2;

   (2)   f   is well defined, in that relabeling the roots

         (e.g. switching m_1 and m_2)   does not alter

         the value of   f.   This is because ϕ is even

         in μ_1 and μ_2,   and is symmetric:

         $\phi(\mu_1,\mu_2) = \phi(\mu_2,\mu_1)$.

   (3)   f   is analytic where (3.5) holds, since m_1 and

         m_2 locally are analytic functions of   z;   and

   (4)   f   is continuous for all   z.

From (3) and (4) we conclude the points $z^4 = 0, \dfrac{q^2}{4}$ are

removable singularities of   f,   and so

                    f   is an entire function of   z.

        Now, to show   f   is in $\mathcal{X}$,   we consider $|z|$ large and

make specific choices

$$m_1 = (\frac{1+i}{\sqrt{2}}) z ((1 - \frac{1}{4} q^2 z^{-4})^{1/2} + \frac{i}{2} qz^{-2})^{1/2}$$

$$= (\frac{1+i}{\sqrt{2}}) z + k_1 (z^{-1})$$

$$m_2 = (\frac{-1+i}{\sqrt{2}}) z ((1 - \frac{1}{4} q^2 z^{-4})^{1/2} - \frac{i}{2} qz^{-2})^{1/2}$$

$$= (\frac{-1+i}{\sqrt{2}}) z + k_2 (z^{-1})$$

of the roots of (3.4). The functions $k_j(w)$ are analytic near $w = 0$, and $k_j(0) = 0$. Therefore,

$$e^{m_1} = \exp((\frac{1+i}{\sqrt{2}}) z) k_3 (z^{-1})$$

where $k_3(w)$ is analytic near $w = 0$ and $k_3(0) = 1$; and similarly for e^{m_2}. A final substitution into (3.6), with a bit of calculation, shows for large $|z|$

$$f(z,q) = z^4 \sum_{k=1}^{5} h_k(z^{-1}, q) e^{c_k z}$$

where $c_1, \ldots, c_4$ are $\pm\sqrt{2}$ and $\pm i\sqrt{2}$, $c_5 = 0$, $h_k(w,q)$ is analytic in w near $w = 0$, and $h_k(0,q) < 0$. Thus $f \in \mathcal{N}$, and the zeros of f (with the possible exception of the five values (3.5)) correspond to the eigenvalues of (3.2), (3.3).

If it were further shown f was a regular homotopy, then roots could be followed as q varies. Even if this were not the case, simple roots can be followed with the assurance

they remain bounded. Alternatively, if the eigenvalues for a

single fixed value of q are sought, an artificial (but

perhaps simpler) homotopy may be appropriate. For example

$$g(z,t) = tf(z,q)$$
$$- (1-t) \prod_{j=1}^{4} (z - a_j)(\cos \gamma z - b)(\cosh \gamma z - c)$$

with $\gamma = (1+i)/\sqrt{2}$ is in $\mathcal{N}$ for almost every a_j, b, and c.

For almost every a_j, b and c, this homotopy is regular.

We remark that the procedure outlined above for obtaining the

analytic function f from the boundary value problem (3.2),

(3.3) is actually quite general, and applies to a variety of

problems.

        One other class of problems we may consider is systems of

reaction-diffusion equations, such as

$$u_t = Du_{xx} + Au$$

where $u(s,x) \in R^n$, $x \in [0,1]$ (say),   D > 0 and A are

matrices; various boundary conditions may be imposed; see [7]

for example. Again, stability questions are of interest. A

function f(z) is obtained, where quite often (generally

depending on the boundary conditions) f $\in$ $\mathcal{F}$. Typically,

$\Pi(f)$ is an interval $[-c,c]$ in which case a homotopy $tf(z) + (1-t)z^p[a \sinh(cz) - b]$ is a regular equivalence for almost every complex a and b.

Finally, we mention that other classes of boundary value problems lead to more complicated situations; these can be studied in a systematic way as above.

REFERENCES

[1] Bellman, R. and Cooke, K.L., Differential-Difference Equations, Academic Press, New York, 1963.

[2] Bolotin, V.V., Nonconservative Problems of the Theory of Elastic Stability, Pergamon Press, New York, 1963.

[3] Carr, J. and Malhardeen, M.Z.M., Beck's Problem, SIAM J. Appl. Math., 37 (1979) 261-262.

[4] Carr, J. and Malhardeen, M.Z.M., Stability of nonconservative linear systems, Functional Differential Equations and Bifurcation (ed. by A.F. Ize) Springer Lecture Notes, vol. 799 (1980), 45-68.

[5] Chow, S.-N., Mallet-Paret, J., and Yorke, J., Finding zeros of maps: homotopy methods that are constructive with probability one, Math. Comp. 32 (1978) 887-899.

[6] Chow, S.-N., Mallet-Paret, J., and Yorke, J., A homotopy method for locating all zeros of a system of polynomials, Functional Differential Equations and Approximation of Fixed Points (ed. by H.-O. Peitgen and H.-O. Walther) Springer Lecture Notes vol. 730 (1979), 77-88.

[7]   Diekmann, O. and Temme, N.M., (eds.), <u>Nonlinear</u> <u>Diffusion</u>
        <u>Problems</u>, vol. 28, Mathematisch Centrum, Amsterdam,
        1976.

[8]   Drexler, F.J., A homotopy method for the calculation of all
        zero-dimensional polynomial ideals, Continuation
        Methods, 69-93, (ed. by H. Wacker), Academic Press,
        New York, 1978.

[9]   Dunford, N. and Schwartz, J.T., <u>Linear</u> <u>Operators</u>, Wiley-
        Interscience, 1971.

[10] Walker, J.A. and Infante, E.F., A perturbation approach
        to the stability of undamped linear elastic systems
        subject to follower forces, J. Math. Anal. Appl. 63
        (1978), 654-677.

WHERE SOLVING FOR STATIONARY POINTS

BY LCPs IS MIXING NEWTON ITERATES

B. Curtis Eaves

Stanford University
Stanford, California 94305

ABSTRACT

A stationary point for a convex polyhedral set and a continuously differentiable function with positive semi-definite derivatives is computed by iteratively solving the linearized problem which
is a linear complementarity problem (LCP). The procedure is shown
to be a mixing of a finite number of Newton methods all converging
to the same points, and consequently, to have convergence properties like Newton's methods.

1. INTRODUCTION

Let X be a set in R^n and $g: X \to R^n$ be a function. A
point x_* in X is defined to be a stationary point of (or solution to the variational problem of) (X, g) if

$$x_*^T g(x_*) \le x^T g(x_*)$$

for all x in X. In this paper we are concerned with the computation of stationary points of (X, g) where $X = \{(u, v): Au + Bv \le c\}$
$\subseteq R^\ell \times R^m = R^n$ and $g(u, v) = (e, f(v))$; the vector e is any element of R^ℓ and the function $f: R^m \to R^m$ is continuously differen-

tiable and has positive definite derivatives.

The algorithm defined below is shown to be a mixing of a finite number of Newton methods all converging to the same point. Our analysis concentrates on the behavior of v as opposed to (u,v); convergence conclusions regarding u follow from those of v. We shall refer to $\bar{v}$ as a v-stationary point of $(X,(e,f))$, if there is a $\bar{u}$ such that $(\bar{u},\bar{v})$ is a stationary point of $(X,(e,f))$. For each $\bar{v}$ we define the function $Lf\bar{v}: R^m \to R^m$ by

$$Lf\bar{v}(v) = f(\bar{v}) + f'(\bar{v})^T(v-\bar{v}).$$

Regard $Lf\bar{v}$ as an affine approximation of f from $\bar{v}$.

Define $G: R^m \to R^m$ by setting $G(\bar{v})$ to be the v-stationary point of $(X,(e,Lf\bar{v}))$, if it exists. The algorithm is defined by $v_k = G^k(v_0)$ for $k = 1,2,\ldots$ for a starting point v_0 in R^m. That is to say, the kth iterate is $v_k = G(v_{k-1})$ where v_{k-1} is the $(k-1)^{th}$ iterate. Executing step k, that is computing $G(v_{k-1})$, is a matter of solving a linear complementarity problem (LCP). Finite procedures for computing $G(v)$ can be found in Cottle[1] and Lemke[7]; however, of special interest here is [2] where advantage can be taken of estimates of $G(v)$.

If v_0 is near a v-stationary point v_* of $(X,(e,f))$, then the iterates v_k generated by the algorithm are shown to converge at a super-linear rate to v_*. If, in addition, the derivative f' of f is Lipschitz continuous at v_*, the convergence rate is quadratic.

As our algorithm is a mixing of a finite number of Newton methods all converging to the same point our results relate closely to those that are concerned with applying Newton's methods or quasi-Newton's methods in the presence of boundaries; see for example, Wilson,[15] Han,[4] Palomares and Mangasarian,[8] Powell,[9,10] and Tapia.[14] Of particular interest is Robinson[11,12,13] wherein stationary points or solutions to "generalized equations" are studied in the presence of perturbations, and these results are used to analyze the limiting

behavior of sequences of solutions of linear complementarity problems; however, his approach required a bounded solution set and our
convergence rate is not obtained there. This paper was initially
motivated by the energy model and algorithms of Hogan[5,6], see [3].

2. PRELIMINARIES

In this section we cite a number of known and/or elementary
results which are necessary for or enhance the subsequent development.

First, let X and Y be sets in R^n and $g: R^n \to R^n$ be a
function on R^n.

Lemma 1. If x is a stationary point of (X,g) and x is
in Y, then x is a stationary point of $(X \cap Y, g)$. If X is
convex, x is a stationary point of $(X \cap Y, g)$, and x is interior
to Y, then x is a stationary point of (X,g). ∎

By a face φ of a convex set X we mean the meet of X with
a supporting hyperplane or the set X itself.

Lemma 2. If X is convex, x is a stationary point of (X,g),
and x lies in the relative interior of a face φ of X, then x
is a stationary point of (Y,g) where Y is the affine hull of φ.

Proof. If y is in Y, then $y = x + \lambda(z-x)$ where z is
in φ and $\lambda > 0$, consequently, $(x-y)^T g(x) = \lambda(x-z)^T g(x) \leq 0$.

For X closed and convex define $\pi: R^n \to X$ to be the projection to X, namely, $\pi(x)$ is the point in X closest to x;
observe that π is continuous. ∎

Lemma 3. Let X be closed and convex. Then x is a stationary point of (X,g), if and only if $\pi(x - g(x)) = x$. If X is
convex and compact and $g: X \to R^n$ is continuous, then (X,g) has
a stationary point.

Proof. $\pi(x - g(x))$ is a continuous map from X to X and
has a fixed point. ∎

Let g and h be functions from X in R^n to R^n. Let $\|g-h\|$ be the supremum of $\|g(x) - h(x)\|$ for x in X where $\|\ \|$ is a norm in R^n.

Lemma 4. If X is compact, then for every $\varepsilon > 0$, there is a $\delta > 0$ such that $\|g-h\| \leq \delta$ implies any fixed point of h lies within ε of some fixed point of g. ∎

For $f: R^m \to R^m$ define f to be monotone or strictly monotone, if

$$(\bar{v} - v)^T(f(\bar{v}) - f(v))$$

is nonnegative or positive for all distinct $\bar{v}$ and v in R^m, respectively; and define f to be strongly monotone, if for some $\alpha > 0$

$$(\bar{v} - v)^T(f(\bar{v}) - f(v)) \geq \alpha\|\bar{v} - v\|^2$$

for all $\bar{v}$ and v in R^m. We say the derivatives of f are positive semi-definite or positive definite, if $y^T f'(v) y$ is nonnegative or positive, respectively, for all (y,v) in $R^m \times R^m$ with $y \neq 0$, and we say the derivatives of f are uniformly positive definite, if for some $\alpha > 0$

$$y^T f'(v) y \geq \alpha\|y\|^2$$

for all (y,v) in $R^m \times R^m$.

Lemma 5. Assuming f is continuously differentiable with positive semi-definite, positive definite, or uniformly positive definite derivatives, then f is monotone, strictly monotone, or strongly monotone, respectively.

Proof. The results follow from the fundamental theorem of integral calculus, that is,

$$f(y) - f(x) = \int_0^1 f'(x + t(y - x))(y - x)dt . \quad ∎$$

Lemma 6. If f is continuously differentiable and strongly monotone, then the derivatives of f are uniformly positive definite.

If f is continuously differentiable and monotone, then the derivatives of f are positive semi-definite. ∎

Let X be a set in $R^{\ell} \times R^{m \cdot}$ with $\ell + m = n$. Let $g(u,v) =$
$(e,f(v))$ where $e \in R^{\ell}$ is fixed and $f: R^m \to R^m$.

Lemma 7. If f is strictly monotone, then $(X,(e,f))$ has at
most one v-stationary point v_*.

Proof. Let (u_0,v_0) and (u_*,v_*) be stationary points. Then

$$(u_* - u_0)^T e + (v_* - v_0)^T f(v_*) \leq 0$$
$$(u_0 - u_*)^T e + (v_0 - v_*)^T f(v_0) \leq 0$$

hence
$$(v_* - v_0)^T (f(v_*) - f(v_0)) \leq 0,$$

and hence $v_0 = v_*$. ∎

For v in R^m define $U(v)$ to be the set of all optimal
solutions u of the program.

$$\text{minimize: } e^T u$$
$$u$$

subject to: $(u,v) \in X$.

Lemma 8. If f is strictly monotone and v_* is the v-stationary point of $(X,(e,f))$, then the set of all stationary points
of $(X,(e,f))$ is $U(v_*) \times \{v_*\}$. ∎

Let X and Y be sets in $R^{\ell} \times R^m = R^n$.

Lemma 9. If f is strictly monotone and some stationary point
of $(Y,(e,f))$ is also a stationary point of $(X,(e,f))$, then all
stationary points of $(Y,(e,f))$ in X are stationary points of
$(X,(e,f))$.

Proof. If (u_1,v_1) and (u_2,v_1) are both stationary points
of $(Y,(e,f))$, then $e^T u_2 = e^T u_1$ and hence, (u_2,v_1) is a stationary point of $(X,(e,f))$. ∎

Let $X = \{(u,v): Au + Bv \le c\} \subseteq R^{\ell} \times R^{m}$, $\quad e \in R$, and $f: R^{m} \to R^{m}$.

Lemma 10. If $(X,(e,f))$ has a stationary point, then

$$Au \le 0 \qquad e^{T}u < 0$$

has no solution. ∎

The next lemma guarantees that each step of our algorithm, namely, the evaluation of $G(v_k)$, can be executed.

Lemma 11. If $Au \le 0$ with $e^{T}u < 0$ has no solution, and $f'(v)$ is positive definite, then $(X,(e,Lfv))$ has a stationary point.

Proof. According to [2] if Lemke's algorithm cannot compute a stationary point of $(X,(e,Lfv))$ then there is a ray $\{(u_1,v_1) + \theta(u_2,v_2): \theta \ge 0\}$ in X with $u_2^{T}e + v_2^{T}(Lfv(v_1+\theta v_2))$ negative, that is, $u_2^{T}e + v_2^{T}f(v) + v_2^{T}f'(v)(v_1-v) + \theta v_2^{T}f'(v)v_2$ negative for all θ. Since $f'(v)$ is positive definite, $v_2 = 0$. Hence, $Au_2 \le 0$ and $u_2^{T}e < 0$ which is contrary to our hypothesis. ∎

Lemma 12. If $Au \le 0$ with $e^{T}u < 0$ has no solution, then there are numbers η and ξ for which $e^{T}u \ge \eta\|v\| + \xi$ for all (u,v) in X.

Proof. Consider the set (u,v,w,z) such that (u,v) is in X, $w = e^{T}u$, $z \ge v_i$, and $z \ge -v_i$ for all i. Use Dines–Fourier–Motzkin elimination to reduce the system to inequalities in two variables w and z. Among these inequalities there must be at least one of form $w \ge \eta z + \xi$. ∎

The next lemma is our best attempt to prove that $(X,(e,f))$ has a stationary point. That X is a polyhedron seems to be of no help; let X be in $R^{\ell} \times R^{m}$.

Lemma 13. If X is closed and convex, if there are numbers η and ξ for which $e^{T}u \ge \eta\|v\| + \xi$ over (u,v) in X, and if f is strongly monotone, then $(X,(e,f))$ has a stationary point.

Proof. Let X_t be the set of x in X such that $\|x\| \leq t$ for $t = 1,2,3,\ldots$. Select $(\bar{u},\bar{v})$ in some X_r. For $t \geq r$ let (u_t,v_t) be a stationary point of $(X_t,(e,f))$. If the v_t's have a cluster point v_* as $t \to +\infty$, then v_* is a v-stationary point of $(X,(e,f))$. Thus, suppose that $v_t \to \infty$ as $t \to +\infty$. We have

$$(u_t - \bar{u})^T e + (v_t - \bar{v})^T f(v_t) \leq 0$$

and hence,

$$\alpha \|v_t - \bar{v}\|^2 \leq (v_t - \bar{v})^T (f(v_t) - f(\bar{v}))$$

$$\leq (\bar{u} - u_t)^T e + (\bar{v} - v_t)^T f(\bar{v})$$

and hence,

$$\|v_t - \bar{v}\| \leq \frac{(\bar{u} - u_t)^T e}{\|v_t - \bar{v}\|} \alpha^{-1} + \|f(\bar{v})\| \alpha^{-1}$$

and hence for all sufficiently large t

$$\|v_t - \bar{v}\| \leq \frac{-u_t^T e}{\|v_t\|} \nu + \rho$$

for some ν and ρ that are invariant with t. By assumption the last expression is bounded and this contradicts our supposition that $v_t \to \infty$. ∎

Let X be a convex polyhedral set in R^n and let $d(x) = d \in R^n$ and $d_*(x) = d_* \in R^n$ be constant functions.

Lemma 14. Assume x_* is a stationary point of (X,d_*). There is an $\varepsilon > 0$ such that if $\|d - d_*\| \leq \varepsilon$ and x is a stationary point of (X,d), then x is in the face $\{y \in X: y^T d_* = x_*^T d_*\}$ of X.

Proof. Suppose the contrary, and we have $d_k \to d_*$ and x_k in X so that $d_*^T x_* < d_*^T x_k$ and $d_k^T x_k \leq d_k^T x$ for all x in X. Select an infinite subsequence so that all x_k lie in the relative interior of one face of X. Letting $\bar{x}$ be any element in the relative interior of this face we have $d_*^T x_* < d_*^T \bar{x}$ and $d_k^T \bar{x} \leq d_k^T x$ for all x in X. Setting $x = x_*$ and taking the limit we get a contradiction. ∎

Let X and Y be sets in R^n. We define the Hausdorf measure of distance between them to be

$$d(X,Y) = \sup_{x \in X} \inf_{y \in Y} \| x - y \|$$

$$+ \sup_{y \in Y} \inf_{x \in X} \| x - y \|$$

where $\| \ \|$ is a norm on R^n.

For the next lemma let $X(c) = \{x: Cx \leq c\} \subseteq R^n$.

Lemma 15. If $X(c_*) \neq \phi$, then there is an η such that $X(c) \neq \phi$ implies

$$d(X(c_*), X(c)) \leq \eta \| c - c_* \| .$$

Proof. An easy proof follows by using induction and Dines-Fourier-Motzkin elimination. ∎

3. CONTINUITY OF G

Given $X = \{(u,v): Au + Bv \leq c\} \subseteq R^\ell \times R^m$, $e \in R^\ell$, and $f: R^m \to R^m$. Define $G: R^m \to R^m$ by setting $G(v)$ to be the v-stationary point of $(X,(e,Lfv))$. We proceed to show G is continuous, if f is continuously differentiable and has positive definite derivatives.

Theorem 1. Let $f: R^m \to R^m$ be continuously differentiable with positive definite derivatives. If G is defined somewhere, then it is well-defined and continuous.

Proof. If (u_1,v_1) is a stationary point of $(X,(e,Lfv_0))$, then $e^T u < 0$ with $Au \leq 0$ has no solution, so G is well-defined by Lemma 11. Select $(\bar{A},\bar{B},\bar{c})$ so that $\bar{X} = \{(u,v) \in X: \bar{A}u + \bar{B}v \leq \bar{c}\}$ is bounded and $\bar{A}u + \bar{B}v < \bar{c}$ for all (u,v) with $\| (u,v) - (u_1,v_1) \| \leq 3\varepsilon$. Select $\varepsilon_1 \leq \varepsilon$ so that $\| v - v_1 \| \leq \varepsilon_1$ implies $d(\bar{U}(v),\bar{U}(v_1)) \leq \varepsilon$ where $\bar{U}(v)$ is the set u that minimizes $e^T u$ subject to (u,v) in $\bar{X}$; see Lemma 15. Using Lemma 4 for $\bar{X}$ and the function

$g(u,v) = \bar{\pi}((u,v) - (e,Lfv_0^{\cdot}(v)))$ where $\bar{\pi}$ is the projection to $\bar{X}$, select δ so that if $\|v_0 - v_2\| \leq \delta$, then each fixed point, say $(\bar{u}_3, v_3)$, of $\bar{\pi}((u,v) - (e,Lfv_2(v)))$ is within ε_1 of some fixed point of $\bar{\pi}((u,v) - (e,Lfv_0(v)))$, say $(\bar{u}_1, v_1)$. Since $\|v_3 - v_1\| \leq \varepsilon_1$ there is a u_3 in $\bar{U}(v_3)$ so that $\|u_3 - u_1\| \leq \varepsilon$; (u_3, v_3) is a stationary point of $(\bar{X}, (e,Lfv_2))$. But $\|(u_3, v_3) - (u_1, v_1)\| \leq 2\varepsilon$ so (u_3, v_3) is a stationary point of $(X, (e,Lfv_2))$ and $\|v_3 - v_1\| \leq \varepsilon.\blacksquare$

4. NEWTON'S METHOD

Let us consider the application of Newton's method to solve the system

$$(1) \qquad\qquad H(u,v) = Eu + F(v) = 0$$

of n equations in n variables (u,v) where Eu is linear in u in R^{ℓ} and F is continuously differentiable in v in R^m.

Given the estimate (u_0, v_0) in $R^{\ell} \times R^m$ the Newton iterate is any (u_1, v_1) in R^n which solves

$$Eu_0 + F(v_0) + E(u_1 - u_0) + F'(v_0)(v_1 - v_0) = 0$$

that is, which solves

$$(2) \qquad\qquad Eu_1 + LFv_0(v_1) = 0 .$$

Note that (u_1, v_1) depends upon v_0 but not u_0 and that such a (u_1, v_1) may not exist. Let us assume that (u_*, v_*) solves (1), that any solution (u,v) to

$$(3) \qquad\qquad Eu + LFv_*(v) = 0$$

has $v = v_*$, and that the Newton iterates (u_k, v_k), $k = 1, 2, \ldots$ can be generated for v_0 sufficiently close to v_*.

Observe, we do not make the customary assumption that the matrix $(E, F'(v_*))$ is nonsingular; however, we have required that $F'(v_*)$ be of full column rank through the uniqueness condition.

We argue that Newton's method in this setting has the usual

convergence properties for v. Namely, first there is a $\delta > 0$ and
$\beta < 1$ such that if $\|v_0 - v_*\| \leq \delta$, then $\|v_1 - v_*\| \leq \beta \|v_0 - v_*\|$.
Second, β can be made arbitrarily small by making δ sufficiently
small. Third, if F' is Lipschitz continuous at v_*, then there
is a γ so that $\|v_1 - v_*\| \leq \gamma \|v_0 - v_*\|^2$ for sufficiently small δ .

    Premultiplying $(E, F'(v_*))$ by a nonsingular matrix Q to
obtain

$$
\begin{array}{c@{\qquad}c}
& \ell \qquad\quad m \\
\begin{array}{c} \overline{\ell} \\[18pt] \overline{m} \end{array} &
\begin{array}{|c|c|}
\hline
E_1 & F_1 \\
\hline
0 & F_2 \\
\hline
\end{array}
\end{array}
$$

where E_1 has linearly independent rows, E_1 is $\overline{\ell} \times \ell$, and
$\overline{\ell} \leq \ell$. Letting

$$
Q = \begin{pmatrix} Q_1 \\ \\ Q_2 \end{pmatrix}
$$

where Q_1 is $\overline{\ell} \times n$, solving (3) is equivalent to solving

(4) $Q_2 LFv_*(v) = 0$

for v and then selecting u so that

(5) $Q_1 Eu = - Q_1 LFv_*(v)$.

Given our assumption that v_* uniquely solves (3), there is an
$m \times m$ submatrix $\overline{Q}_2$ of Q_2 so that $\overline{Q}_2 F'(v_*)$ is square and non-
singular. Thus, solving (3) solves

(6) $\overline{Q}_2 LFv_*(v) = 0$.

    Now by classical results, upon applying Newton's method to
solve
(7) $\overline{Q}_2 F(v) = 0$

there is a $\delta > 0$ and $\beta < 1$ so that if $\|v_0 - v_*\| \leq \delta$, then

$\|v_1 - v_*\| \leq \beta \|v_0 - v_*\|$.   Second, β can be made arbitrarily small
by making δ sufficiently small. Further, if $\overline{Q}_2 F'$ is Lipschitz
continuous at v_*, then there is a γ so that $\|v_1 - v_*\| \leq \gamma \|v_0 - v_*\|^2$
for sufficiently small δ.

Given v_0, let $v_1, v_2, \ldots$ be the sequence of v_k's generated
by applying Newton's method to (1). This same sequence is generated
by applying Newton's method to (7) and thus our argument is complete.

In view of Lemma 15, we have $d(U(v_k), U(v_*)) \leq \eta \|v_k - v_*\|$ for
some η and $k = 1, 2, \ldots$.

5. THE FUNCTIONS G_i AND MIXING

Let $X = \{(u,v): Au + Bv \leq c\} \subseteq R^\ell \times R^m$, $e \in R$, and $f: R^m \to R^m$
be continuously differentiable with positive definite derivatives.
Let φ_i be any closed face of X of any dimension and let X_i be
the affine hull of φ_i.

For v in R^m define $G_i(v)$ to be the v-stationary point of
$(X_i, (e, Lfv))$, if it exists; recall that if $(X_i, (e, Lfv))$ has a
v-stationary point, then it is unique; see Lemma 7. We proceed to
show that the action of G_i is equivalent to that of Newton's method
as in the previous section.

Define $H_i: R^\ell \times R^m \to R^n$ by $H_i(u,v) = \pi_i((u,v) - (e, f(v))) - (u,v)$ where π_i is the projection to X_i. π_i is an affine function and can be expressed in the form $\pi_i(x) = Px + p$. Hence, H_i
is linear in u. Given (u_0, v_0) the Newton iterate of H_i is any
solution (u_1, v_1) in R^n to

$$H_i(u_0, v_0) + H_i'(u_0, v_0)((u_1, v_1) - (u_0, v_0)) = 0$$

or

$$P((u_0, v_0) - (e, f(v_0))) + p - (u_0, v_0)$$

$$+ \left(P - P\left(\begin{array}{c|c} 0 & 0 \\ \hline 0 & f'(v_0) \end{array}\right) - I\right)((u_1, v_1) - (u_0, v_0)) = 0$$

or

$$P((u_1, v_1) - (e, Lfv_0(v_1))) + p - (u_1, v_1) = 0$$

or

$$\pi_i(u_1,v_1) - (e,Lfv_0(v_1))) - (u_1,v_1) = 0 .$$

That is to say, given v_0, the Newton iterate of H_i is any (u_1,v_1), which is a stationary point of $(X_i,(e,Lfv_0))$. That is, v_1 of the Newton iterate (u_1,v_1) is $G_i(v_0)$.

Given v_0 let (u_1,v_1) be a stationary point of $(X,(e,Lfv_0))$, that is, $v_1 = G(v_0)$. Then (u_1,v_1) is interior to some face φ_i of X, thus,

$$G(v_0) = G_i(v_0) = v_1 .$$

Repeating this observation for (u_1,v_1), (u_2,v_2), etc. we see that G is a mixing of the G_i's or that G is a mixing of a finite number of Newton methods.

Now suppose that (u_*,v_*) is a statonary point of $(X_i,(e,f))$. The v-stationary point v_* of $(X_i,(e,f))$ is unique, and by Lemmas 10 and 11 the iterates of G_i can be generated. Thus, from the previous section there is a $\delta_i > 0$ and $\beta_i < 1$ such that if $\|v_0 - v_*\| \le \delta_i$, then $\|G_i(v_0) - v_*\| \le \beta_i\|v_0 - v_*\|$. Second, β_i can be made arbitrarily small by making δ sufficiently small. Third, if f is Lipschitz continuous at v_*, there is a γ_i and $\delta_i > 0$ so that for $\|v_0 - v_*\| \le \delta_i$, $\|G_i(v_0) - v_*\| \le \gamma_i\|v_0 - v_*\|^2 .$

6.   LOCAL CONVERGENCE RATE OF G

Take X, e, and f as in the previous section. Let v_* be the v-stationary point of $(X,(e,f))$. We show that if v_0 is sufficiently near v_*, then $G^k(v_0)$ tends to v_* at a superlinear rate, and if, in addition, f' is Lipschitz continuous, the rate is quadratic.

Select any u_* in $U(v_*)$ and let φ_1 be the face $\{(u,v) \in X: (e,f(v_*))^T((u_*,v_*) - (u,v)) = 0\}$ of X and let $\varphi_1,\varphi_2,\dots,\varphi_h$ be all faces of φ_1 that meet $(U(v_*),v_*)$. Of course, φ_1 contains $U(v_*) \times \{v_*\}$. Defining X_i and G_i for $i = 1,\dots,h$ as in the previous section, we observe that any point in $\varphi_i \cap (U(v_*),v_*)$ is

a stationary point of $(X_i, (e,f))$ for $i = 1,\ldots,h$. According to
the previous section, select $\delta = (\min \delta_i) > 0$ and $\beta = (\max \beta_i) < 1$
so that for $\|v_0 - v_*\| \le \delta$ we have $\|G_i(v_0) - v_*\| \le \beta \|v_0 - v_*\|$ for
$i = 1,\ldots,h$. The factor β can be made arbitrarily small by making
δ sufficiently small. If f' is Lipschitz continuous at v_*, then
there is a γ so that

$$\|G_i(v_0) - v_*\| \le \gamma \|v_0 - v_*\|^2$$

for sufficiently small δ for $i = 1,\ldots,h$.

Select $\Delta_1 \le \delta$ so that if (u,v) is in φ_1 and within a
distance Δ_1 of the set $U(v_*) \times \{v_*\}$, then (u,v) is in the
relative interior of one of the faces $\varphi_1,\ldots,\varphi_h$. Select $\Delta_2 \le (1/2)\Delta_1$
so that for $\|v - v_*\| \le \delta_2$ we have $d(U(v), U(v_*)) \le (1/2)\Delta_1$. Select
$\varepsilon > 0$ according to Lemma 14 for $d_* = (e, f(v_*))$ and φ_1. Select
$\Delta_3 \le \Delta_2$ so that for v_0 and v within a distance Δ_3 of v_* we
have $\|Lfv_0(v) - f(v_*)\| \le \varepsilon$. Select $\Delta_4 \le \Delta_3$ so that for $\|v_0 - v_*\| \le \Delta_4$
we have $\|G(v_0) - G(v_*)\| \le \Delta_3$.

Now for $\|v_0 - v_*\| \le \Delta_4$ we have $\|G(v_0) - v_*\| \le \Delta_3$. Consequently,
$\|Lfv_0(G(v_0)) - f(v_*)\| \le \varepsilon$. Selecting any u_1 in $U(G(v_0))$ we have
$(u_1, G(v_0))$ in φ_1. Since $\|G(v_0) - v_*\| \le \Delta_2$ there is a point u_*
in $U(v_*)$ with $\|(u_1, G(v_0)) - (u_*, v_*)\| \le \Delta_1$. Consequently,
$(u_1, G(v_0))$ lies interior to one of the faces $\varphi_1,\ldots,\varphi_h$. There-
fore, $G(v_0) = G_i(v_0)$ for some $i = 1,\ldots,h$. Thus, $\|G(v_0) - v_*\| \le$
$\beta \|v_0 - v_*\|$. Reusing the last expression, we obtain $G^k(v_0) \to v_*$.
Since β can be made arbitrarily small by making δ sufficiently
small we obtain $\|G^{k+1}(v_0) - v_*\| \le \beta_k \|G^k(v_0) - v_*\|$ where $\beta_k \to 0$ as
$k \to \infty$. Similarly, if f' is Lipschitz continuous at v_*, we have

$$\|G^{k+1}(v_0) - v_*\| \le \gamma \|G^k(v_0) - v_*\|^2$$

for $k \ge K$ for some K.

In view of Lemma 15, we have $d(U(v_k), U(v_*)) \le \eta \|v_k - v_*\|$ for
some η for $k = 1,2,\ldots$. If $U(v_*) \times \{v_*\}$ is interior to φ_1,
then $G_1(v_0) = G(v_0)$ for v_0 near v_*, and we see that G is

Newton's method and not merely a mixing of such. If $U(v_*) \times \{v_*\}$
is equal to φ_1, we see that $G^k(v_0) = v_*$ for some k for v_0
near v_*, that is, the algorithm converges finitely.

7. REMARKS

As Newton's algorithm is not globally convergent one cannot
expect the algorithm of the paper to be globally convergent, that
is, let $\ell = 0$, $m = 1$, and $X = [0,1]$.

The algorithm of this paper has not been programmed and/or
tested.

REFERENCES

1. R. W. Cottle, The principal pivoting method of quadratic pro-
      gramming, in "Mathematics of the Decision Sciences, Part I,"
      G. B. Dantzig and A. F. Veinott, Jr., eds., Am. Math. Soc.:
      144-162 (1968).
2. B. C. Eaves, Computing stationary points, Math. Prog., Study 7:
      1-14 (1978).
3. B. C. Eaves, A locally quadratically convergent algorithm for
      computing stationary points, Dept. of Operations Research,
      Stanford University (May 1978).
4. S.-P. Han, Superlinearly convergent variable metric algorithms
      for general nonlinear programming problems, Math. Prog. 11:
      263-282 (1976).
5. W. W. Hogan, Energy policy models for project independence,
      Comput. & Ops. Res. 2: 251-271 (1975).
6. W. W. Hogan, Project independence evaluation system: structure
      and algorithms, Proc. of Symposia in Appl. Math. 21 (1977).
7. C. E. Lemke, Bimatrix equilibrium points and mathematical pro-
      gramming, Management Sciences 11: 681-689 (1965).
8. U. M. G. Palomares and O. L. Mangasarian, Superlinearly conver-
      gent quasi-Newton algorithms for nonlinearly constrained
      optimization problems, Math. Prog. 11: 1-13 (1976).
9. M. J. D. Powell, A fast algorithm for nonlinearly constrained
      optimization calculations, Dept. of Appl. Math. and Theor.
      Physics, Univ. of Cambridge, England (June 25, 1977).
10. M. J. D. Powell, The convergence of variable metric methods
      for nonlinearly constrained optimization calculations, Dept.
      of Appl. Math. and Theor. Physics, Univ. of Cambridge,
      England (August 5, 1977).
11. S. M. Robinson, An implicit-function theory for generalized
      variational inequalities, MRC TSR #1672, Univ. of Wisconsin,
      Madison (September 1976).

12. S. M. Robinson, Generalized equations and their soluctions, part I: Basic theory, Math. Prog., Study 10: 128-141 (1979).

13. S. M. Robinson, Generalized equations and their solutions, Part II: Applications to nonlinear programming, MRC TSR #2048, University of Wisconsin, Madison (March 1980).

14. R. A. Tapia, Diagonalized multiplier methods and quasi-Newton methods for constrained optimization, Optimization Theory and Applications 22: 135-194 (June 1977).

15. R. B. Wilson, A simplicial algorithm for concave programming, D.B.A. Thesis, Harvard University (May 1963, Rev'd May 1964).

Other related readings include:

16. B. C. Eaves, On the basic theory of complementarity, Math. Prog. 1: 68-75 (October 1971).

17. P. E. Gill and W. Murray, Newton-type methods for unconstrained and linearly constrained optimization, Math. Prog. 7: 311-350 (1974).

18. L. Mathiesen, Computational experience from solving nonlinear complementarity problems by Newton-like process, Dept. of Operations Research, Stanford University (October 1981).

19. J. S. Pang and D. Chan, Iterative methods for variational and complementarity problems, Graduate School of Industrial Administration, Carnegie-Mellon University, Pittsburgh (April 1981).

20. M. J. Todd, A note on computing equilibria in economics with with activity analysis models of production, Jr. of Math. Econ. 6: 135-144 (1979).

ON THE EQUIVALENCE OF THE LINEAR COMPLEMENTARITY PROBLEM AND

A SYSTEM OF PIECEWISE LINEAR EQUATIONS: PART II

B. C. Eaves* and C. E. Lemke**

*Department of Operations Research, Stanford University,
Stanford, California 94305

**Department of Mathematical Sciences, Rensselaer
Polytechnic Institute, Troy, New York 12181

ABSTRACT

In an earlier paper the authors demonstrated the equivalence
of the linear complementarity problem and that of finding the zeroes
of a square system of equations for which the functions are piecewise linear. Given the system of equations, a "dual" concept of
complementarity was evoked to pose the problem as an LCP.

In this paper, with the simplifying assumption of a nondegenerate finite sub-division defined by hyperplanes, a simplified
equivalence is demonstrated, wherein "complementarity" refers only
to the positive vs negative sides of a hyperplane.

KEY WORDS: Piecewise linear, linear complementarity, subdivision
by hyperplanes, Non-degenerate.

INTRODUCTION

We consider as the Linear Complementarity Problem that of finding solutions z, w to:

[1] $0 = q + Mw + Nz$; w, $z \geq 0$; and $z^T w = 0$;

where M and N are given n x n square matrices. (This may be put into the usual form: $w = q + Az$ when there exists a non-singular complementary matrix B from (-M,N); i.e., for each i, column B_i is either $-M_i$ or N_i.)

In the first paper on the subject (Ref. [2]) the authors noted that the LCP [1] is readily posed as that of finding zeroes of a piecewise linear system, (see also Ref. [3]). We record this now and discuss it, for use as an integral part of the paper.

For a column, say x, the positive and negative parts of x, which we label x^+ and x^-, are the unique columns satisfying:

[2] $x = x^+ - x^-$; x^+, $x^- \geq 0$; and $x^{+T} x^- = 0$.

(Thus, given q, $(w, z) = (q^+, q^-)$ is the unique solution to the simplest LCP: $w = q + z$; w, $z \geq 0$; $w^T z = 0$.)

With this notation, given LCP [1], upon defining:

[3] $f(x) := q + Mx^+ + Nx^-$,

then $f(x) = 0$ iff $(w, z) = (x^+, x^-)$ solves LCP. f is piecewise linear on the (simple) subdivision defined by the n hyperplanes with equations $x_i = 0$; $i = 1, 2, \ldots, n$. In this example, the cells of the subdivision are the 2^n orthants of "x-space", R^n.

The remainder of this paper is concerned with going the other way. Briefly, we shall suppose as given some k hyperplanes in x-space, described as follows. Let (b, A) be given, where A is kxn and no row of A is the zero row.

Consider the equations:

[4] $y := b + Ax$.

Thus, y_i is an affine function of x. Let H_i^o denote the set of x for which $y_i = 0$. Let H_i^+ denote the set of all x for which $y_i \geq 0$; that is, for which $y_i^- = 0$; that is, for which $y_i = y_i^+$; and let H_i^- denote the set for which $y_i \leq 0$. H_i^o is a hyperplane, with corresponding positive halfspace H_i^+, and negative halfspace H_i^-. Note that y_i^+ and y_i^- are also functions of x. We will consider the subdivision of R^n by the k hyperplanes H_i^o.

In this introduction, we develop a statement of the main theorem. In the next section we discuss the geometry of subdivisions by hyperplanes, and then all functions piecewise linear (PL) on such a subdivision, and prove the main theorem. In a final section, note is made of a more general setting which was brought to our attention (Ref. [5]), by A. Polyméris.

To begin, consider again the "LCP" function f in [3]. It has n real components, f_i, each PL over the simple "orthant" subdivision. It will suffice in all of our discussion to limit consideration to real-valued functions. Thus, a function $f: R^n \longrightarrow R$, given by:

[5] $f(x) := c + dx^+ + ex^-$; where (c, d, e) is 1x(2n+1)

is PL over the orthant subdivision, called S temporarily. Consider all such functions of the form [5]. Clearly they form a vector space of dimension 2n+1, equal to the number of arbitrary elements in the row (c, d, e). Clearly, also, the set of all functions PL over S also forms a vector space. It will emerge as a special case that any such PL function over S has the form [5]. The form [5] is a special case of the following: given B, n x n and non-singular, consider the set of all functions of the form:

[6] $f(x) = c + dy^+ + ey^-$,

where:

[7] $y := b + Bx$.

Again, f is PL over the subdivision defined by the "independent" hyperplanes given by $y_i = 0$, $i = 1, 2, \ldots, n$. Note, in particular, that the "ramp" function expressed by $f(x) := y_i^+$ is PL over the subdivision, since it is 0 (thus affine) on all cells of the subdivision where $y_i = y_i^-$, and is linear $(f(x) = y_i)$ in the region where $y_i = y_i(x) = y_i^+$. More generally consider our subdivision of interest so that now y as in [7] is as given in [4]. We shall show that, when A has rank n, any function PL over the subdivision takes the form [6]. However, we state the theorem using possibly a more illuminating form, as follows: writing $y = y^+ - y^-$, solving for y^- and substituting $y^- = y^+ - y = y^+ - b - Ax$ in [6] we obtain functions $f: R^n \dashrightarrow R$ of the form:

[8] $f(x) = g(x) + h(y^+);$

where $g: R^n \dashrightarrow R$ affine and $h: R^n \dashrightarrow R$ linear. Thus, $g(x) := c' + d'x$, and $f(x) - g(x)$ is a sum of ramp functions. The main result is that any function PL over the subdivision takes this form, which is slightly more general than that in [6].

SUBDIVISIONS BY HYPERPLANES.   MAIN THEOREM.

    We shall write: $<k> := \{1, 2, \ldots, k\}$ and consider subsets K of $<k>$. We assume that the k hyperplanes H_i^o are distinct. We write S(K) to denote the subdivision of R^n by the subset of hyperplanes H_i^o, i in K, but for K = $<k>$ we write S := S($<k>$).

    Given K := $\{i_1, i_2, \ldots, i_r\}$ we define the set:

[9] $H(K) := H_{i_1}^o \cap H_{i_2}^o \cap \ldots \cap H_{i_r}^o$.

For a general K, H(K) might be empty. We next identify the cells of S. Consider the 2^k convex polyhedra of the form:

[10] $H_1^{t_1} \cap H_2^{t_2} \cap \ldots \cap H_k^{t_k}$; where each t_i is "+" or "-" .

Each of these is a cell C of S iff it is n-dimensional. The cells comprise S, and their union is R^n. More generally, if K $\subseteq$ K*, each cell of S(K) is a union of cells of S(K*) and we say that S(K*) is "finer" than S(K) (S(K) "coarser"). Thus, S is finer than S(K). Consider K for which H(K) is not empty. Note that the cells of

S(K) are "cones based on H(K)" (a ray from a point in H(K) lies totally in a cell of S(K)). We shall call such an S(K) a "conical subdivision".

DEFINITION: Consider p in R^n. Star p is defined as the union of all cells of S which contain p.

LEMMA 1. Star p is the intersection of all half spaces containing p. In particular, consider p in R^n. Let K(p) denote the set of all indices of hyperplanes containing p. If we set: $K^* := \langle k \rangle - K(p)$, then note that Star p is a cell of $S(K^*)$, hence is convex. □

DEFINITION: Let P denote a convex r-dimensional subset of R^n. We define the "restriction of S to P", denoted $S|P$, to be the set of all r-dimensional sets of the form $C \cap P$, where C is a cell of S. Thus, the elements of $S|P$ subdivide P. Similarly for $S(K)|P$. We are interested mainly in r=n or n-1. If r=n, we will consider P = Star p, or P = N(p), a neighborhood of point p. Consider a hyperplane, H^o and non-empty K. Thus, $S(K)|H^o$ is a subdivision of H^o. In particular, consider an H_i^o, where i is in $\langle k \rangle$. Then the cells of $S|H_i^o$ are "facets" of S. Conversely, a facet of S is a set $F := C \cap H_i^o$, for some i in $\langle k \rangle$ and some cell C of S when F is (n-1)-dimensional.

If C is a cell of S and p is in Int C, note that K(p) is empty, and Star p = C.

DEFINITION: We say that $f: R^n \longrightarrow R$ is "Piecewise-linear on S(K)", denoted S(K)-PL, iff f is affine on every cell of S(K).

Note that if f is S(K)-PL, then f is also $S(K^*)$-PL whenever $S(K^*)$ is finer than S(K), since cells of S(K) are unions of cells of $S(K^*)$. In particular, if f is affine, it is S(K)-PL for any K. By definition, if $f: R^n \longrightarrow R$ is S(K)-PL, and C is a cell of S(K), there is an affine function $g: R^n \longrightarrow R$, such that f equals g on C. A more general instance of this is given in Lemma 2:

LEMMA 2. Given p, let K := K(p). Let $f: R^n \longrightarrow R$ be S-PL. Then f restricted to Star p may be extended to a function $g^*: R^n \longrightarrow R$ which is S(K)-PL.

Proof: Each cell C of S(K) is conical from H(K). C∩Star p is a

cell of S. The affine function $g:R^n$ ----> R which agrees with f

on C∩Star p is also affine in C. All such g's over all cells C of

S(K) identify a g* which is S(K)-PL. □

Let $f:R^n$ ----> R be affine. It is characteristic of affine

functions that f is completely determined when its value is known

at any affinely independent set of n+1 points (vertices of an

n-simplex).

LEMMA 3. Let f be S-PL, and p a point. If f is affine in a

neighborhood of p, then f is affine on Star p.

Proof: Let K* := <k>-K(p). Star p is a cell, call it C^*, of $S(K^*)$,

and p is in Int C*. Consider a neighborhood N*(p), wholly in

Int C*. If f is affine in N*(p). it is determined by the values on

the vertices of any n-simplex contained in N*(p). Hence f is affine

on cell C*=Star p. □

DEFINITION: In R^n, a hyperplane H^o "separates" a set T iff there

are points of T in both Int H^+ and Int H^-.

LEMMA 4. In R^n let P denote an n-dimensional convex set, T denote

a convex subset of P, and let H^o be a hyperplane that separates T.

Consider the (two-cell) subdivision of R^n by H^o. Let f:P ----> R

be PL with respect to this subdivision. If f is affine on T, then

f is affine on P.

Proof: If $g:R^n$ ----> R is the affine function which agrees with

f on $P∩H^+$; if p is in T∩Int H^+, and q is in T∩Int H^-, take n

affinely independent points on H^o. These together with p determine

g everywhere, in particular in T and in particular at q. Then the

n points on H^o and point q determine f in $P∩H^-$, which is therefore

g. □

THEOREM 1. Consider K, where 1 is in K. Suppose that H^o_1 separates

H(K-1). Let f be S(K)-PL. If f is affine in a neighborhood N(p)

for some point p in H, then f is S(K-1)-PL.

Proof: Let C denote a cell of S(K-1) containing p. Referring to

Lemma 4, let P: = C; T: = H(K-1) and H^o: = H_1^o. Applying Lemma 4, f

is affine on C.   Hence f is S(K-1) - PL. □

DEFINITION: If S is a subdivision by hyperplanes, then S is "non-

degenerate" iff for any K ⊂ <k>, if H(K) is non-empty, then the set

of hyperplanes corresponding to indices in H(K) is independent

(that is, the set of normals to the hyperplanes is an independent

set of vectors).

        We are now in a position to prove (see, also, Addendum):

THEOREM 2.   Let   S :=S(<k>) be a non-degenerate subdivision of

x-space, defined by y = b + Ax.   Let f be S-PL.   Then there is an

affine map $g:R^n$ ----> R, and a linear map $h:R^n$ ----> R such that:

[11]   f(x) = g(x) + h(y^+).

Proof: We proceed by induction on k.   For   k=1 define g by

g(x) = f(x), for x in H_1^-.   Thus, f(x) = g(x)+(f(x)-g(x)) for all x.

The function f(x)-g(x) is zero in H_1^- and may be expressed as

f(x)-g(x)=:h(y_1^+), where h(y_1) :=ay_1, some a.

        Next we assume the assertion true for all r < k, k ≥ 2,

and prove it for k.   Select a point p meeting H_1^o but not H_i^o for

i = 2, ..., k.   As done for the case k=1, select g*(x) and h*(y_1)

so that:

[12]   f*(x) := f(x) - [g*(x)+h*(y_1^+)]

is zero in a neighborhood of p, not meeting H_i^o, i=2, 3, ...,k.

        Define:

[13]   Q := {q in H_1^o: for some N(q), f* is PL on S(<k>-1)|N(q)}.

Clearly, Q is open in H_1^o and p is in Q, so that Q ≠ ∅ .   Hence,

if we show that Q is closed, then Q = H_1^o.   For this, let <p_j>

denote a sequence of points converging to p*, say.   Consider K(p*).

If K(p*) = {1}, so that p* is in Int F for some facet F of S,

then surely p* is in Q, since any neighborhood of p* contains points

of Q.   Otherwise, |K(p*)| > 1, and then, since S is non-degenerate

(see NOTE at end of Proof), H_1^o separates H(K(p*)-1 and Theorem 1 may be applied in the proof. In any case, $K(p*)$ corresponds to an independent set of hyperplanes. There is a neighborhood $N(p*)$ which meets only these hyperplanes. Let $f**:R^n$ ----> R denote the $S(K(p*))$-PL function which agrees with $f*$ on $N(p*)$, hence on Star $p*$ (see Lemma 2).

Now for j large enough there is a neighborhhod of p_j, say $N(p_j)$, wholly contained in $N(p*)$, and $N(p_j)$ satisfies the require-ment of Q, and meets no hyperplanes other than those which meet $p*$. Thus, $f**$, hence $f*$ is $S(<k>-1)$-PL in $N(p_j)$, (and henceforth we ignore $f**$).

We may take a point q in $N(p*)$ which meets only H_1^o. Hence q is in Int.F, where F is a facet of S, meeting cells C^+ in H_1^+ and C^- in H_1^- of S. Thus, $C^+\cup C^-$ =: C is a cell of $S(<k>-1)$. Hence, $f*$ is affine on C, and so, by Theorem 1, since H_1^o separates $H(K(p*)-1)$, $f*$ is $S(K(p*)-1)$-PL. Therefore, $f*$ is $S(<k>-1)$-PL, since $S(<k>-1)$ is finer. In particular, $f*$ is $S(<k>-1)$-PL in $N(p*)$, so that $p*$ is in Q. It follows that Q is closed in H_1^o, and hence $Q=H_1^o$.

It follows that $f*$ is affine in every cell of $S(<k>-1)$, so that $f*$ is $S(<k>-1)$-PL.

By the induction hypothesis, we therefore may write:

$$f*(x) = g**(x) + h**(y^{*+}); \quad (y* \text{ is } y \text{ with } y_1 \text{ deleted}),$$

where $g**$ is affine and $h**$ is linear. Thus, the Theorem is proved by setting:

$$g = g* + g**; \text{ and } h = h* + h**. \quad \square$$

NOTE: As noted in the proof, the use of "non-degenerate" was limited to the situation where necessarily H_1^o separated $H(K-1)$, where $K = K(p*)$. The following example illustrates a degenerate sub-division in R^2. The PL function indicated cannot be expressed in the form [11]. Note that the points where f is given are 7 in number -- the vector space of all functions S-PL has dimension 7, greater than $(n+1) + k=6$, the number of independent coefficients

in [11]. The set of all functions of form [11] is a 6-dimensional
subspace.

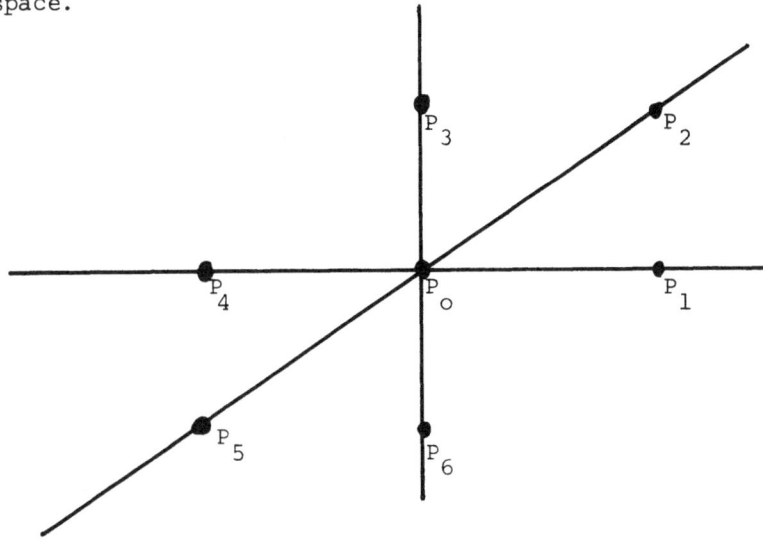

Figure 1

CORROLLARY: Given a non-degenerate subdivision of R^n by k hyperplanes, let V denote the vector space of all functions PL over the
subdivision. Then:

[14]   Dim. V = (n+1)+k.

    Now consider the general case of a non-degenerate subdivision
by hyperplanes as in [4], which we repeat here as:

[15]   y :=b + Ax;   where A is k x n. By Theorem 2, any
function PL over this subdivision, S, has the form

[16]   f(x) : (a+Cx) + Dy^+, where C is n x n and D is n x k. We
now suppose that S has a vertex (then every cell has a vertex),
which is the case iff Rank A = n. Thus, some n independent hyperplanes meet, and for some permutation matrix, call it P:

[17] $Py =: \begin{pmatrix} y^1 \\ y^2 \end{pmatrix} = P(b+Ax) =: \begin{pmatrix} b^1 \\ b^2 \end{pmatrix} + \begin{pmatrix} B \\ E \end{pmatrix} x,$

where B is non-singular, and we may solve for x, giving:

[18] $x = B^{-1}(y^1 - b^1),$

so that y=b+Ax is given in terms of y by:

[19] $0 = -y^2 + b^2 + Ex = -y^2 + b^2 + EB^{-1}(y^1-b^1)$

         $= G(y-b);$ where $G := (EB^{-1}, -I)P,$ (k-n) x k.

Likewise we may replace x in [16] to obtain:

[20] $f(x) = a + CB^{-1}(y^1-b) + Dy^+$

         $= (a-CB^{-1}b^1) + (CB^{-1}, 0)Py + Dy^+$

         $= (a-CB^{-1}b^1) + [(CB^{-1}, 0)P+D]y^+ + [-(CB^{-1}, 0)P]y^-$

         $= q + M y^+ + N y^-,$ where M and N are n x k.

Now, combining [20] and [19] into a single equation yields:

[21] $\begin{pmatrix} f(x) \\ 0 \end{pmatrix} = \begin{pmatrix} q \\ -Gb \end{pmatrix} + \begin{pmatrix} M \\ G \end{pmatrix} y^+ + \begin{pmatrix} N \\ -G \end{pmatrix} y^-,$ and $\begin{pmatrix} M \\ G \end{pmatrix}, \begin{pmatrix} N \\ -G \end{pmatrix}$ are each k×k.

so that finding zeroes of f(x) is the same as solving an LCP. We record this fact:

CORROLLARY. Let S = S(<k>) be a non-degenerate subdivision of R^n with a vertex. If f is S-PL, then finding zeroes of f(x) is the same as finding all solutions to some LCP.

A GENERAL SETTING. FINAL COMMENTS.

     With reference to Figure 1, using the characteristic of affine function that it is completely determined by its values at the n+1 vertices of any n-simplex, we noted that the values at the particular 7 points shown suffice to determine an S-PL function uniquely. We concluded that the dimension of V, the vector space of all functions S-PL, was seven. Indeed, clearly, V is the set of all linear combinations of the particular 7 S-PL functions f, i = 0, 1, ..., 6, defined by the values $f_i(P_i) = 1, f_i(P_j) = 0, j \neq i.$

     In fact, we may refer to the set of the seven points as an "S-basis", since that term depends only on S. More generally, let T denote a non-empty set and consider the vector space:

[22] $R^T := \{f: \quad f:T \longrightarrow R\}$.

Let V denote a fixed subspace of R^T of finite dimension d.

Let:

[23] $B := \{g_1, g_2, \ldots, g_d\}$

denote a fixed basis of V. If x is in T write:

[24] $B(x) := (g_1(x), g_2(x), \ldots, g_d(x))$ in R^d.

DEFINITION: Let $X \subseteq T$. We say that X is "V-independent" iff the set:

[25] $B(X) := \{B(x) : x \text{ in } X\}$

is linearly independent in R^d.

Thus, if X is V-independent, $|X| \leq d$. As observed by C. W. Lee (Ref. [4]), X is V-independent iff, for all $f^*:X \longrightarrow R$, there exists f in V such that $f|X = f^*$.

As a classical example, if T :=R, and

V := $\{f:R \longrightarrow R : f$ a polynomial of degree $\leq r\}$,

then Dim V = r+1.

For the example of the paper, if S denotes a (finite) sub-division of R^n; $T:R^n$, and V is the space of all functions S-PL, V has finite dimension, which is n+1+k when S is a non-degenerate subdivision by k hyperplanes. A "pre-geometry" (or "matroid") discussion of this is found, for example, in Ref. [1].

As a final remark, and summary note, we have shown that, when S is a non-degenerate subdivision by some k hyperplanes then any f S-PL may be exhibited in the "closed form" [11] of Theorem 2, namely as the sum of a simple affine function, and simple "ramp" functions. The general case (not necessarily non-degenerate) appears to be more involved, as the example of Figure 1 indicates and will be analyzed in a future paper. In this regard, the "matroid" approach alluded to in Ref. [1] appears promising.

ADDENDUM

Our main theorem, namely Theorem 2, is a special case, with respect to "splines of degree 1", of the principal result in Zwart [6]; nevertheless, our proof is different and has merits of its own.

The authors would like to thank Michael J. Todd for bringing Zwart's paper to their attention.

REFERENCES

[1] Crapo, H. H., and G.-C. Rota, On the Foundations of Combinatorial Theory: Combinatorial Geometries (Preliminary Edition), The MIT Press, 1970.

[2] Eaves, B. C., and C. E. Lemke, "Equivalence of LCP and PLS", Mathematics of Operations Research, Vol. 6, No. 4, November 1981.

[3] Eaves, B. C., and Herbert Scarf, "The Solution of Systems of Piecewise Linear Equations", Mathematics of Operations Research, Vol. 1, No. 1, February 1976.

[4] Lee, Carl W., Private Communication, Department of Mathematical Sciences, IBM Research Laboratory, Yorktown Heights, N.Y., September 1980 - September 1981.

[5] Polymeris, A., Private Communication, IFOR, ETH, Zurich.

[6] Zwart, Philip B., " Multivariate Splines with Nondegenerate Partitions", SIAM Journal of Numerical Analysis, Vol. 10, No. 4, September 1973.

RELATIONS BETWEEN PL MAPS, COMPLEMENTARY

CONES, AND DEGREE IN LINEAR COMPLEMENTARITY PROBLEMS[*]

C. B. Garcia[1], F. J. Gould[2], and T. R. Turnbull

Graduate School of Business

University of Chicago

1. Introduction

Let M be an $n \times n$ real matrix and q an n-vector. The problem

$$\text{find } z \in R^n \text{ and } w \in R^n \text{ such that}$$

$(I, -M, q)$ $\qquad w = q + Mz, \quad w \geq 0, \quad z \geq 0, \quad \text{and}$

$$w^T z = 0 \text{ (i.e., } w_i = 0 \text{ or } z_i = 0 \text{ for } i = 1, \ldots, n)$$

is called the linear complementarity problem (LCP). This problem is a canonical form for a variety of significant problems in math-

---

[*]The authors are indebted to R. W. Cottle and R. E. Stone for insightful discussions and comments on an earlier version of this paper.

[1]The research of this author was partially supported by NSF Grant #MCS 77-15509.

[2]The research of this author was partially supported by NSF Grant #ECS 79-20177 and by the Centre de Recherche de Mathematiques de la Decision of the University of Paris IX. This author also acknowledges with gratitude several fundamental and technical insights provided by T. Dittmer.

ematical programming, economics and engineering (e.g., references

[2], [6], [9], [18-20], [24], [32]) and there is a wealth of

literature on the problem (see, for example, references [3-5],

[7], [10-17], [21-23], [25-28], [30-31], [33-43] and papers cited

therein).

In most previous literature this problem has been treated

from an algorithmic/algebraic point of view, as initiated most

notably by Lemke [18] and Cottle and Dantzig [2]. However a small

but increasingly large body of work has approached the problem

from a geometric point of view, beginning notably with papers of

Murty [24], Saigal [34-35], and then with recent work of Kelly and

Watson [13] and Cottle and von Randow [4]. In these papers the

structure of so-called complementary cones has been studied.

Index theory was first introduced for the complementarity problem

by Saigal and Simon [36]. In recent geometric work Eaves and

Scarf [7], Megiddo and Kojima [23], Kojima and Saigal [15-16], and

Howe [10] have begun to study the LCP from the viewpoint of PL

(piecewise linear) maps and the topological theory of degree. The

purpose of the present paper is to pursue these latter courses of

development. Specifically the complementary cone approach is

unified with the PL approach and the theory of degree is heavily

used to obtain several insights. Thus, the work herein is heavily

motivated by the approaches of Murty, Kelly and Watson, Cottle,

Eaves and Scarf, and Howe. The geometric degree-oriented approach

of the paper leads to the construction of a new PL homotopy

algorithm for the LCP, which is presented in [8]. More
specifically, the paper is organized as follows.

In Section II some of the basic notation is recorded.

Section III discusses background material on relations
between PL maps, complementary cones, and the LCP. It is a modest
attempt to unify some structure which is already known but
recently known, and it lays out the geometric framework to be
employed throughout. The notions of weak nondegeneracy and degree
are reviewed. A concept called "conical degree" is introduced.
It will be employed in the remaining Section IV.

In Section IV attention is focused on Q-matrices, those
square matrices for which the linear complementarity problem has a
solution for any $q \in R^n$. Numerous researchers have already
attacked the problem of characterizing such matrices and various
sets of necessary and sufficient conditions are known. But the
problem of obtaining a generally considered "useful characterization" remains unsolved. In Section IV several results are added
to the existing lore. Specifically,

1. In R^n, $n \leq 3$, a strongly nondegenerate matrix M
(i.e., we shall speak of matrices which are weakly nondegenerate,
nondegenerate, and strongly nondegenerate) is a Q-matrix if and
only if the conical degree of any associated PL map is nonzero
(Proposition 4.2 and Theorem 4.6).

2. It is shown that a nondegenerate Q-matrix M can have a
zero conical degree. This disposes of a previously conjectured

possible characterization of Q-matrices (see, for example, [10]).

3. A theoretic characterization of Q is given in

Proposition 4.1.

4. Q-matrices are shown to be related to lower dimensional

Q-matrices.

5. A new conjecture for an interesting characterization of

Q is given. The conjecture is proven to be true in R^n, for n =

2 or 3 (Theorem 4.7).

II. Notation

Let A and B denote two real n × n matrices. This pair

of matrices (A, B) is said to be nondegenerate if the columns

A_1, ..., A_n, B_1, ..., B_n have the property that every sequence of

vectors ξ_1, ..., ξ_n, where ξ_i is either A_i or B_i, is a

linearly independent set (in particular, the matrices A and B

are nonsingular). Let $<\xi_1, ..., \xi_n>$ denote the closed convex

polyhedral cone in R^n spanned by a given choice of n ξ_i's,

i.e., the set of all linear combinations of the ξ_i with

nonnegative coefficients. The 2^n (not necessarily distinct)

cones spanned by the 2^n such sequences of ξ_i's are called the

complementary cones spanned by (determined by, generated by, etc.)

the matrices A and B (complementary by virtue of the fact that

each generator ξ_i is an A_i or a B_i). The lower dimensional

faces of these complementary cones are also complementary cones.

For example, the (closed) face $<\xi_{j_1}, ..., \xi_{j_k}>$ is a cone, and is called a complementary k-face, or a complementary k-cone. The set

of all complementary cones containing this k-face are the 2^{n-k}

complementary extensions of this face, denoted by

$C<\xi_{j_1}, \ldots, \xi_{j_k}>$. For example, suppose $n = 3$. The complementary

extensions of the 1-face $<A_2>$ are $C<A_2> = \{<A_1, A_2, A_3>,$

$<B_1, A_2, A_3>, <A_1, A_2, B_3>, <B_1, A_2, B_3>\}$. The complementary

(n-1)-faces are called complementary facets.

Let $N = \{1, 2, \ldots, n\}$ and let K be a naturally ordered

subset of N. It is understood that $K = \phi$ is allowed. The

symbol $|K|$ is the cardinality of K, and $\overline{K}$ will denote the

naturally ordered complement of K in N. Given a real $n \times n$

matrix M, with entries m_{ij}, the principal submatrix with

entries m_{ij}, $i \in K$, $j \in K$, will be denoted M_{KK}. The columns

of a matrix M are denoted M_i, and for $\phi \neq K \subset N$, $\phi \neq J \subset N$,

the notation M_{KJ} denotes the $|K| \times |J|$ submatrix whose

elements are m_{ij}, $i \in K$, $j \in J$. If $z \in R^n$ and $K \subset N$ we at

times use the notation z_K to denote the $|K|$-dimensional vector

with components z_i for $i \in K$. For example, if

$N = \{1, \ldots, 6\}$ and $K = \{1, 3, 5\}$, then z_K is the three-

dimensional vector $\begin{bmatrix} z_1 \\ z_3 \\ z_5 \end{bmatrix}$. For a matrix α, α_K denotes the

submatrix whose columns are indexed by K. The determinant of the

matrix M is denoted $\det M$ and the determinants of the

principal submatrices, $\det M_{KK}$ for $K \subset N$, are the principal

minors of M. By assumption, $\det M_{\phi\phi} = 1$.

For an   n × n   matrix   M,

$$\| M \| = \left(\Sigma_{i=1}^{n} \Sigma_{j=1}^{n} m_{ij}^{2} \right)^{1/2} ,$$

the euclidean norm in $R^{n \times n}$,   and

$$\| M \|_{L} = \sup \{ \| Mx \| : x \in R^{n}, \quad \| x \| = 1 \} ,$$

the norm of   M   viewed as a linear transformation from R^{n} into R^{n}.

It is understood throughout that all quantities are real.

III. Background

A. PL Maps and the LCP

It will be useful to review some known relations between the LCP and certain PL maps, and to review some facts about the degree of such maps.

Given any nonsingular matrix   α,   note that the pair (α, -α)   is nondegenerate in the sense of Section II (the complementary cones spanned by   (α, -α)   are all of dimension n). Moreover the 2^{n} possible closed complementary cones spanned by this pair are distinct and they partition R^{n} in the sense that they cover the entire space and no two cones have overlapping interiors.

It will be convenient to represent the complementary cones spanned by   (α, -α)   with the following notation. Corresponding to each subset   K ⊂ N   is the closed complementary cone (generated by $(\alpha, -\alpha))$

$$\overline{C}_K(\alpha) = \{x \in R^n : x = \sum_{i \in \overline{K}} \lambda_i \alpha_i - \sum_{i \in K} \lambda_i \alpha_i, \ \lambda_i \geq 0, \ i = 1, \ \ldots, \ n\}$$

$$= \langle \alpha_{\overline{K}}, -\alpha_K \rangle \ .$$

Let $C_K(\alpha)$ denote its interior. From time to time it will be convenient to refer to the union of the boundary facets of the cones $\overline{C}_K(\alpha)$, $K \subset N$. We denote this closed set of measure zero in R^n as $H(\alpha)$.

Now let A and B be any pair of $n \times n$ matrices (not necessarily nondegenerate). The four matrices α, $-\alpha$, A, and B (or, more briefly, the three matrices α, A, B) provide the structure for a PL map $f_{\alpha, A, B}$ as follows (cf. Eaves and Scarf [7], Megiddo and Kojima [23], and Howe [10]). For each $x \in R^n$, there is a $K \subset N$ such that $x \in \overline{C}_K(\alpha)$. Then there are unique, nonnegative numbers $\lambda_1(x), \ \ldots, \ \lambda_n(x)$ such that

$$x = \sum_{i \in \overline{K}} \lambda_i(x)\alpha_i - \sum_{i \in K} \lambda_i(x)\alpha_i = [\alpha_{\overline{K}}, -\alpha_K] \begin{bmatrix} \lambda_{\overline{K}}(x) \\ \lambda_K(x) \end{bmatrix}$$

and hence

$$\begin{bmatrix} \lambda_{\overline{K}}(x) \\ \lambda_K(x) \end{bmatrix} = [\alpha_{\overline{K}}, -\alpha_K]^{-1} x \ .$$

Now define

$$f_{\alpha, A, B}(x) = \sum_{i \in \overline{K}} \lambda_i(x)A_i + \sum_{i \in K} \lambda_i(x)B_i = [A_{\overline{K}}, B_K] \begin{bmatrix} \lambda_{\overline{K}}(x) \\ \lambda_K(x) \end{bmatrix}$$

$$= [A_{\overline{K}}, B_K] [\alpha_{\overline{K}}, -\alpha_K]^{-1} x$$

for $x \in \overline{C}_K(\alpha)$.

If $x \in \overline{C}_K(\alpha) \cap \overline{C}_J(\alpha)$ for $K \subset N$ and $J \subset N$, then $\lambda_i(x) = 0$ for $i \in (K \cap \overline{J}) \cup (\overline{K} \cap J)$. Hence

$$\sum_{i \in \overline{K}} \lambda_i(x) A_i + \sum_{i \in K} \lambda_i(x) B_i = \sum_{i \in \overline{K} \cap \overline{J}} \lambda_i(x) A_i + \sum_{i \in K \cap J} \lambda_i(x) B_i$$

$$= \sum_{i \in \overline{J}} \lambda_i(x) A_i + \sum_{i \in J} \lambda_i(x) B_i$$

so that $f_{\alpha, A, B}$ is well-defined on R^n. The range of this map will be denoted by $R(\alpha, A, B)$.

In order to simplify notation, the PL map of interest will often be denoted simply by f, with the underlying structure (α, A, B) understood. Let $f_K: \overline{C}_K(\alpha) \to R^n$ be the restriction of f to $\overline{C}_K(\alpha)$, i.e., $f_K \triangleq f|_{\overline{C}_K(\alpha)}$. (The n-dimensional vector $f_K(x)$ should not be confused with the $|K|$-dimensional vector $(f(x))_K$ whose components are $f_i(x)$ for $i \in K$.) Since f_K is linear, f is continuous on each closed cone $\overline{C}_K(\alpha)$. Because there are only finitely many of the $\overline{C}_K(\alpha)$, f is continuous on R^n.

Let Lf_K denote the linear extension of f_K to R^n. Since f_K is linear, f is differentiable on each open cone $C_K(\alpha)$ and, for any $x \in C_K(\alpha)$,

$$f'(x) = f'_K(x) = Lf'_K(x) = [A_{\overline{K}}, B_K] [\alpha_{\overline{K}}, -\alpha_K]^{-1} .$$

Note that

$$f(\overline{C}_K(\alpha)) = f(<\alpha_{\overline{K}}, -\alpha_K>) = Lf_K(<\alpha_{\overline{K}}, -\alpha_K>) = <A_{\overline{K}}, B_K>$$

so that $R(\alpha, A, B)$ equals the union of the complementary cones

spanned by (A, B).

We will also use the notation

$$\eta_{K,i} = \begin{cases} \alpha_i, & i \in \overline{K} \\ \\ -\alpha_i, & i \in K \end{cases}$$

and

$$\xi_{K,i} = \begin{cases} A_i, & i \in \overline{K} \\ \\ B_i, & i \in K \end{cases}$$

For $x \in \overline{C}_K(\alpha)$,

$$x = \sum_{i=1}^{n} \lambda_i(x)\eta_{K,i} = [\eta_{K,1}, \cdots, \eta_{K,n}] \begin{bmatrix} \lambda_1(x) \\ \vdots \\ \lambda_n(x) \end{bmatrix}$$

and

$$f(x) = \sum_{i=1}^{n} \lambda_i(x)\xi_{K,i} = [\xi_{K,1}, \cdots, \xi_{K,n}] \begin{bmatrix} \lambda_1(x) \\ \vdots \\ \lambda_n(x) \end{bmatrix}$$

$$= [\xi_{K,1}, \cdots, \xi_{K,n}][\eta_{K,1}, \cdots, \eta_{K,n}]^{-1}x .$$

Let us observe that $\det f_K'(x) \neq 0$ if and only if $[A_{\overline{K}}, B_K]$

is nonsingular. Hence

det $f'_K \neq 0$ for all $K \subseteq N \iff (A, B)$ is nondegenerate .

With this relationship in mind, let us define the map $f_{\alpha,A,B}$ to be nondegenerate if and only if (A, B) is nondegenerate, which is true if and only if det $f'_K \neq 0$ for all $K \subseteq N$. Note that this is equivalent to stating that each f_K is one-to-one on $\overline{C}_K(\alpha)$, or each Lf_K is one-to-one on R^n, or that for each $K \subset N$

$$Lf_K(x) = 0 \iff x = 0$$

(for each $K \subset N$, the null space of Lf_K intersects R^n uniquely at the origin).

Where possible, it will be of interest to weaken this property as follows. The map $f_{\alpha,A,B}$ is said to be weakly nondegenerate if and only if for each $K \subset N$

$$x \in \overline{C}_K(\alpha) \text{ and } Lf_K(x) = 0 \iff x = 0$$

(for each $K \subset N$, the null space of Lf_K intersects $\overline{C}_K(\alpha)$ uniquely at the origin). Since $x \in \overline{C}_K(\alpha)$ implies

$$Lf_K(x) = f_K(x) = f(x) ,$$

it follows that f is weakly nondegenerate if and only if

$$f(x) = 0 \iff x = 0 .$$

Clearly if f is nondegenerate, then f is weakly nondegenerate. It is also to be noted that neither of these conditions guarantees that all of the complementary cones spanned by (A, B) are distinct since, for example, some A_i and B_i may be linearly dependent. However, this possibility will be of no concern.

Throughout this paper the image cones $<A_{\overline{K}}, B_K>$ will be of particular interest along with sgn det f'_K. This latter value may be loosely referred to as "the sign of $<A_{\overline{K}}, B_K>$," and these signs are said to form "the sign pattern" of the image cones.

We shall now illustrate a close connection between PL equations and linear complementarity.

Given $x \in R^n$, consider again the representation

$$x = [\alpha_{\overline{K}}, -\alpha_K] \begin{bmatrix} \lambda_{\overline{K}}(x) \\ \\ \lambda_K(x) \end{bmatrix}, \quad \lambda_{\overline{K}}(x) \geq 0, \quad \lambda_K(x) \geq 0 ,$$

where K is any subset of N such that $x \in \overline{C}_K(\alpha)$. Now define $w(x)$ and $z(x)$ to be the trivial extensions of $\lambda_{\overline{K}}(x)$ and $\lambda_K(x)$ to R^n. That is, for $i = 1, \ldots, n$

$$w_i(x) = \begin{cases} \lambda_i(x), & i \in \overline{K} \\ \\ 0, & i \in K \end{cases}$$

$$z_i(x) = \begin{cases} 0, & i \in \overline{K} \\ \\ \lambda_i(x), & i \in K \end{cases} .$$

Then $w(x) \geq 0$, $z(x) \geq 0$, $w(x)^T z(x) = 0$, and using $w(x)$ and

$z(x)$ we can rewrite the PL map as

$$f_{\alpha,A,B}(x) = Aw(x) + Bz(x) .$$

Using this form for the function f, it is easy to see that

$f_{\alpha,A,B}(x) = v$ implies $w(x)$ and $z(x)$ solve (A, B, v).

Conversely, if w and z solve (A, B, v), then $f_{\alpha,A,B}(x) = v$

for $x = \alpha w - \alpha z$.

With a proper choice of A, B, and v the above correspon-

dence is useful in solving a given LCP $(I, -M, q)$. Given the

matrix M, we shall often require that $f_{\alpha,A,B}$ (or, equivalent-

ly, A and B) satisfies

Assumption $(*)_M$

(i) A is nonsingular

(ii) $B = -\rho AM$ for some positive scalar ρ.

Note that assumption $(*)_M$ is independent of α. Clearly if

$f_{\alpha,A,B}$ satisfies $(*)_M$ and $f_{\alpha,A,B}(x) = Aq$, then $w(x)$ and

$\rho z(x)$ solve $(I, -M, q)$. Conversely, if w and z solve

$(I, -M, q)$ and $f_{\alpha,A,B}$ is any PL map satisfying $(*)_M$, then

$f_{\alpha,A,B}(x) = Aq$ for $x = \alpha w - \alpha z/\rho$.

If $f_{\alpha,A,B}$ satisfies $(*)_M$, the determinants of the $f_K^{\vee}$

are related to the principal minors of M by

$$\det f_K^{\vee} = \frac{\rho^{|K|} \det A}{\det \alpha} \det M_{KK}$$

for all $K \subset N$. It follows immediately that $f_{\alpha,A,B}$ is nondegenerate if and only if all the principal minors of M are nonzero (in which case M is also called <u>nondegenerate</u>).

We define a matrix M to be <u>weakly nondegenerate</u> if, for all $K \subset N$,

$$M_{KK}z_K = 0$$

$$M_{\overline{K}K}z_K \geq 0$$

$$z_K \geq 0$$

imply $z_K = 0$, for any $z \in R^n$. This terminology is justified by

<u>Proposition 3.1.</u> For any PL map $f_{\alpha,A,B}$ satisfying $(*)_M$, $f_{\alpha,A,B}$ is weakly nondegenerate if and only if M is weakly nondegenerate.

<u>Proof.</u> Assume f is weakly nondegenerate. Fix $K \subset N$. Suppose $M_{KK}z_K = 0$, $M_{\overline{K}K}z_K \geq 0$, and $z_K \geq 0$ for some $z \in R^n$. Then

$$\begin{bmatrix} I_{\overline{K}\overline{K}} & -\rho M_{\overline{K}K} \\ 0 & -\rho M_{KK} \end{bmatrix} \begin{bmatrix} M_{\overline{K}K}z_K \\ \frac{1}{\rho}z_K \end{bmatrix} = \begin{bmatrix} M_{\overline{K}K}z_K - M_{\overline{K}K}z_K \\ -M_{KK}z_K \end{bmatrix} = 0 \ .$$

Let

$$x = [\alpha_{\overline{K}}, \ -\alpha_K] \begin{bmatrix} M_{\overline{K}K}z_K \\ \frac{1}{\rho}z_K \end{bmatrix} \ .$$

Since $M_{\overline{K}K}z_K \geq 0$ and $\frac{1}{\rho}z_K \geq 0$, $x \in \overline{C}_K(\alpha)$ and

$$f(x) = [A_{\overline{K}}, \ B_K] \begin{bmatrix} M_{\overline{K}K} z_K \\ \frac{1}{\rho} z_K \end{bmatrix} = [A_{\overline{K}}, \ A_K] \begin{bmatrix} I_{\overline{K}\overline{K}} & -\rho M_{\overline{K}K} \\ 0 & -\rho M_{KK} \end{bmatrix} \begin{bmatrix} M_{\overline{K}K} z_K \\ \frac{1}{\rho} z_K \end{bmatrix}$$

$$= 0 .$$

Hence $x = 0$ by the weak nondegeneracy of f. Therefore

$$\begin{bmatrix} M_{\overline{K}K} z_K \\ \frac{1}{\rho} z_K \end{bmatrix} = [\alpha_{\overline{K}}, \ -\alpha_K]^{-1} x = 0$$

and so $z_K = 0$.

   Conversely, assume that M is weakly nondegenerate. Suppose $f(x) = 0$ for some $x \in R^n$. There is a $K \subset N$ such that $x \in \overline{C}_K(\alpha)$. Then

$$f(x) = [A_{\overline{K}}, \ A_K] \begin{bmatrix} I_{\overline{K}\overline{K}} & -\rho M_{\overline{K}K} \\ 0 & -\rho M_{KK} \end{bmatrix} \begin{bmatrix} \lambda_{\overline{K}}(x) \\ \lambda_K(x) \end{bmatrix} = 0 .$$

Let $z_K = \rho \lambda_K(x)$. Since A is nonsingular, we have

$$\begin{bmatrix} I_{\overline{K}\overline{K}} & -\rho M_{\overline{K}K} \\ 0 & -\rho M_{KK} \end{bmatrix} \begin{bmatrix} \lambda_{\overline{K}}(x) \\ \frac{1}{\rho} z_K \end{bmatrix} = \begin{bmatrix} \lambda_{\overline{K}}(x) - M_{\overline{K}K} z_K \\ -M_{KK} z_K \end{bmatrix} = 0 .$$

Hence $M_{KK} z_K = 0$, $M_{\overline{K}K} z_K = \lambda_{\overline{K}}(x) \geq 0$, and $z_K = \rho \lambda_K(x) \geq 0$. It follows from the weak nondegeneracy of M that $z_K = 0$. Therefore $\lambda_K(x) = 0$ and $\lambda_{\overline{K}}(x) = 0$ so that $x = 0$.   □

   The concept of weak nondegeneracy for a matrix M has been used frequently in linear complementarity theory (see, e.g.,

Saigal [34]). The above proposition shows that this is equivalent

to the concept of weak nondegeneracy for $f_{\alpha,A,B}$ described by

Howe [10], which is, in fact, a necessary condition for the defi-

nition of the conical degree of $f_{\alpha,A,B}$ given in Part B below.

We now illustrate two specific choices of α, A, and B

(i.e., two realizations of the PL function f) such that $(*)_M$

is satisfied.

Example 1. Take $\alpha = I$, $A = I$, $B = -M$. In this case $(*)_M$ is

clearly satisfied with $\rho = 1$. Note that the cone

$$\overline{C}_K(I) = \{x \in R^n: x_i \geq 0, i \in K; x_i \leq 0, i \in K\} = \langle I_{\overline{K}}, -I_K \rangle,$$

which is one of the 2^n closed orthants of R^n. Also it is

easily verified that, for $i = 1, \ldots, n$,

$$w_i(x) = \max\{x_i, 0\} = (|x_i| + x_i)/2$$

$$z_i(x) = \max\{-x_i, 0\} = -\min\{x_i, 0\} = (|x_i| - x_i)/2$$

and
$$x = w(x) - z(x)$$

$$f(x) = f_{I,I,-M}(x) = Iw(x) - Mz(x) = (|x| + x)/2 - M(|x| - x)/2$$

where $|x| = (|x_1|, |x_2|, \ldots, |x_n|) \in R^n$. For $x \in \overline{C}_K(I)$,

$$f(x) = [I_{\bar{K}}, \; -M_K] \begin{bmatrix} \lambda_{\bar{K}}(x) \\ \\ \lambda_K(x) \end{bmatrix} = [I_{\bar{K}}, \; -M_K] \begin{bmatrix} x_{\bar{K}} \\ \\ -x_K \end{bmatrix}$$

$$= [I_{\bar{K}}, \; M_K] \begin{bmatrix} x_{\bar{K}} \\ \\ x_K \end{bmatrix},$$

hence

$$f_K^{\smile} = [\hat{\xi}_{K,1}, \; \cdots, \; \hat{\xi}_{K,n}] \; ,$$

where

$$\hat{\xi}_{K,i} = \begin{cases} I_i, & i \in \bar{K} \\ \\ M_i, & i \in K \end{cases} .$$

Note that

$$\det f_K^{\smile} = \det M_{KK}$$

for all $K \subseteq N$.   If   A   and   B   satisfy $(*)_M$,   then

$$f_{\alpha, A, B} = Af_{I, I, -\rho M} \alpha^{-1} \; .$$

The range of the map $f_{I, I, -M}$, namely $R(I, I, -M)$, is the union of the complementary cones generated by $(I, -M)$. That is, $f(\overline{C}_K(I)) = \langle I_{\bar{K}}, -M_K \rangle$. It follows that $R(I, I, -M)$ is precisely the set of all q such that $(I, -M, q)$ is solvable. These cones generated by $(I, -M)$ have been notably studied by Murty [24], Kelly and Watson [13], and Saigal [34-35] in previous works. The associated function $f_{I, I, -M}$ has been discussed in previous literature by Eaves and Scarf [7], Megiddo and Kojima [23], and Howe [10].

Example 2. Take $\alpha = (M - \mu I)^{-1}$, $A = \mu\alpha$, $B = -M\alpha$, where μ is any positive scalar which is not an eigenvalue of M. In this case, $(*)_M$ is satisfied with $\rho = 1/\mu$. That is, noting that $M = \alpha^{-1} + \mu I$, we have

$$-\rho AM = -\alpha M = -\alpha(\alpha^{-1} + \mu I)$$

$$= -I - \mu\alpha = -(\alpha^{-1} + \mu I)\alpha$$

$$= -M\alpha = B .$$

Hence, $(*)_M$ is satisfied.

We have the following simple representations of $f_{\alpha,\mu\alpha,-M\alpha}$.

Proposition 3.2. For any $K \subseteq N$,

$$\overline{C}_K(\alpha) = \{x \in R^n : (Mx)_i - \mu x_i \geq 0 \text{ for } i \in \overline{K} \text{ and}$$
$$(Mx)_i - \mu x_i \leq 0 \text{ for } i \in K\} .$$

If $x \in \overline{C}_K(\alpha)$, then

$$\begin{bmatrix} (f_{\alpha,\mu\alpha,-M\alpha}(x))_{\overline{K}} \\ (f_{\alpha,\mu\alpha,-M\alpha}(x))_K \end{bmatrix} = \begin{bmatrix} \mu I_{\overline{K}\overline{K}} & 0 \\ M_{K\overline{K}} & M_{KK} \end{bmatrix} \begin{bmatrix} x_{\overline{K}} \\ x_K \end{bmatrix} .$$

Hence, for any $x \in R^n$,

$$(f_{\alpha,\mu\alpha,-M\alpha}(x))_i = \min\{(Mx)_i, \mu x_i\}$$

for $i = 1, \ldots, n$.

Proof. For any $x \in R^n$,

$$\begin{bmatrix} \lambda_{\bar{K}}(x) \\ \lambda_K(x) \end{bmatrix} = \begin{bmatrix} \alpha_{\bar{K}\bar{K}} & -\alpha_{\bar{K}K} \\ \alpha_{K\bar{K}} & -\alpha_{KK} \end{bmatrix}^{-1} \begin{bmatrix} x_{\bar{K}} \\ x_K \end{bmatrix}$$

$$= \begin{bmatrix} M_{\bar{K}\bar{K}} - \mu I_{\bar{K}\bar{K}} & M_{\bar{K}K} - \mu I_{\bar{K}K} \\ -(M_{K\bar{K}} - \mu I_{K\bar{K}}) & -(M_{KK} - \mu I_{KK}) \end{bmatrix} \begin{bmatrix} x_{\bar{K}} \\ x_K \end{bmatrix}$$

$$= \begin{bmatrix} (Mx)_{\bar{K}} - \mu x_{\bar{K}} \\ -((Mx)_K - \mu x_K) \end{bmatrix} .$$

Hence $x \in \bar{C}_K(\alpha)$ if and only if $(Mx)_i - \mu x_i \geq 0$ for $i \in \bar{K}$
and $(Mx)_i - \mu x_i \leq 0$ for $i \in K$.

Note that

$$\alpha M = \alpha(\alpha^{-1} + \mu I) = I + \mu\alpha = (\alpha^{-1} + \mu I)\alpha = M\alpha .$$

For any $x \in \bar{C}_K(\alpha)$,

$$
\begin{bmatrix} (f(x))_{\overline{K}} \\ (f(x))_{K} \end{bmatrix} = \begin{bmatrix} \mu\alpha_{\overline{KK}} & -(M\alpha)_{\overline{KK}} \\ \mu\alpha_{K\overline{K}} & -(M\alpha)_{KK} \end{bmatrix} \begin{bmatrix} \alpha_{\overline{KK}} & -\alpha_{\overline{K}K} \\ \alpha_{K\overline{K}} & -\alpha_{KK} \end{bmatrix}^{-1} \begin{bmatrix} x_{\overline{K}} \\ x_{K} \end{bmatrix}
$$

$$
= \begin{bmatrix} \mu\alpha_{\overline{KK}} & -\mu\alpha_{\overline{K}K} \\ \mu\alpha_{K\overline{K}} & -I_{KK} - \mu\alpha_{KK} \end{bmatrix} \begin{bmatrix} M_{\overline{KK}} - \mu I_{\overline{KK}} & M_{\overline{K}K} \\ -M_{K\overline{K}} & -M_{KK} + \mu I_{KK} \end{bmatrix} \begin{bmatrix} x_{\overline{K}} \\ x_{K} \end{bmatrix}
$$

$$
= \begin{bmatrix} \mu(\alpha M)_{\overline{KK}} - \mu^2\alpha_{\overline{KK}} & \mu(\alpha M)_{\overline{K}K} - \mu^2\alpha_{\overline{K}K} \\ \mu(\alpha M)_{K\overline{K}} - \mu^2\alpha_{K\overline{K}} + M_{K\overline{K}} & \mu(\alpha M)_{KK} - \mu^2\alpha_{KK} - \mu I_{KK} + M_{KK} \end{bmatrix} \begin{bmatrix} x_{\overline{K}} \\ x_{K} \end{bmatrix}
$$

$$
= \begin{bmatrix} \mu I_{\overline{KK}} & 0 \\ M_{K\overline{K}} & M_{KK} \end{bmatrix} \begin{bmatrix} x_{\overline{K}} \\ x_{K} \end{bmatrix} .
$$

Thus

$$
f_i(x) = \begin{cases} \mu x_i, & i \in \overline{K} \\ (Mx)_i, & i \in K \end{cases}
$$

for $x \in \overline{C}_K(\alpha)$. It follows that for any $x \in R^n$

$$
f_i(x) = \min\{(Mx)_i, \mu x_i\}
$$

for $i = 1, \ldots, n$, using the representation for $\overline{C}_K(\alpha)$ obtained above.   □

The PL map $f_{\alpha,\mu\alpha,-M\alpha}$ corresponding to the choice $\alpha = (M - \mu I)^{-1}$ appears not to have been previously studied in the

LCP context. The authors [8] have developed an algorithm for the
LCP based on this map.

Note that for $f = f_{\alpha, \mu\alpha, -M\alpha}$,

$$\det \hat{f}_K = \mu^{|\overline{K}|} \det M_{KK}.$$

Also $(I, -M, q)$ has a solution if and only if $\mu\alpha q \in R(\alpha, \mu\alpha, -M\alpha)$.

The LCP $(I, -M, q)$ has a solution for all q (i.e., M is
a Q-matrix) if and only if the map $f_{\alpha, A, B}$ is surjective, where
$f_{\alpha, A, B}$ is any PL map satisfying assumption $(*)_M$. Thus the
question of whether or not M is a Q-matrix can be studied by
investigating conditions under which these PL maps are surjective,
and this will be taken up in Section IV. However, this topic
leads to considerations involving the degree of these PL maps,
which we now review.

B. The Conical Degree. Let f now denote a continuous PL map
defined by given matrices α, A, B. On the cone $\overline{C}_K(\alpha)$ the
function f is given by a linear expression f_K. Let us begin by
applying the usual notion of degree to the continuous map f. Let
D be a bounded open set in R^n. For any $y \notin f(\partial D)$ the usual
degree theory defines an integer, dependent upon the choice of
y, called the degree of f at y relative to D, and denoted
$\deg(f, D, y)$.

We shall now define a structure which for our purposes will
be more useful than the classical notion of degree. First of all,

let us assume that the PL map f satisfies the <u>weak nondegeneracy</u> condition. That is

$$f(z) = 0 \iff z = 0 .$$

Now let D be any bounded open set containing the origin. Then by the weak nondegeneracy assumption the point $y = 0$ is not in $f(\partial D)$. Thus $\deg(f, D, 0)$ is well defined. We shall define the <u>conical degree of</u> f, denoted $cd(f)$, by

$$cd(f) \overset{\Delta}{=} \deg(f, D, 0) .$$

Thus, $cd(f)$ will be an integer associated with f, independent of D. The following result shows that $cd(f)$ is well-defined and can be calculated by examining the pre-images of any point $y \in R^n - f(H(\alpha))$.

<u>Proposition 3.3.</u> If $f_{\alpha,A,B}$ is weakly nondegenerate and $y \in R^n$ such that $y \notin f(H(\alpha))$ and $y \in R(\alpha, A, B)$, then

$$cd(f) = \sum_{x \in f^{-1}(y)} \text{sgn det } f'(x) .$$

If f is not surjective, then $cd(f) = 0$.

<u>Proof.</u> Let $y \in R^n$ and D a bounded open set containing 0. Since $\deg(f, D, 0)$ is locally constant (see Ortega and Rheinboldt [29]), there is an $\varepsilon > 0$ such that $\|z\| \leq \varepsilon$ implies $z \notin f(\partial D)$ and

$$\deg(f, D, 0) = \deg(f, D, z) \; .$$

If $f^{-1}(y) = \phi$, then $f^{-1}(\varepsilon y/\|y\|) = \phi$ and

$$cd(f) = \deg(f, D, 0) = \deg(f, D, \varepsilon y/\|y\|) = 0 \; .$$

Assume $f^{-1}(y) \neq \phi$ and $y \notin f(H(\alpha))$. It follows that $\det f'(x) \neq 0$ and $x \notin H(\alpha)$ for all $x \in f^{-1}(y)$. Then $f^{-1}(y)$ contains at most 2^n points (no more than one in each $C_K(\alpha)$). For each $x \in f^{-1}(y)$ there is a $\gamma(x) > 0$ such that $\gamma(x)x \in D$. Let

$$\gamma = \min\{\varepsilon/\|y\|, \min\{\gamma(x): x \in f^{-1}(y)\}\}.$$

Since γx and x are in the same open cone,

$$f'(\gamma x) = f'(x) \; .$$

Moreover,

$$f^{-1}(\gamma y) = \gamma f^{-1}(y) \subset D$$

so that $\gamma y \in R^n - f(H(\alpha)) - f(\partial D)$. Hence

$$cd(f) = \deg(f, D, 0)$$

$$= \deg(f, D, \gamma y)$$

$$= \sum_{x \in f^{-1}(\gamma y) \cap D} \operatorname{sgn} \det f'(x)$$

$$= \sum_{x \in f^{-1}(y)} \operatorname{sgn} \det f'(x) \; . \qquad \Box$$

Howe [10] gives an alternate development of these ideas.

Two cones $\overline{C}_K(\alpha)$ and $\overline{C}_J(\alpha)$ are <u>adjacent</u> if they have a common facet. A facet which separates its adjacent cones (i.e., the images of its adjacent cones have nonempty interiors whose intersection is empty) will be called a <u>separating facet</u>. Otherwise it is termed a <u>nonseparating facet</u>. Two adjacent cones $\overline{C}_K(\alpha)$ and $\overline{C}_J(\alpha)$ have a separating facet in common if and only if

$$\text{sgn det } f'_K = \text{sgn det } f'_J \neq 0$$

(see Murty [24] and Saigal [35]).

IV. Results on Q-Matrices

Recall that a Q-matrix is a matrix $M \in R^{n \times n}$ such that $(I, -M, q)$ has a solution for every $q \in R^n$. The set of Q-matrices will often be referred to simply as Q. A number of authors have sought to obtain either interesting or useful descriptions of the class Q (see, for example, references [1], [4-6], [12-13], [31]). While conditions which are either separately necessary or sufficient are known, a satisfactory characterization is yet to be obtained. In this section we dispose of at least one interesting conjecture (raised by several workers, for example, in the paper of Howe [10]) that a non-degenerate matrix M is in Q if and only if the conical degree of an associated PL map is nonzero. Another approach is initiated which leads to the previously unknown consequence that a Q-matrix

is closely related to lower dimensional Q-matrices. Finally a new

conjecture for a characterization of Q is given.

Before giving the main results of this section we recall from

Section III that if the PL map $f_{\alpha, A, B}$ satisfies assumption

$(*)_M$, i.e.,

(i) A is nonsingular

(ii) $B = -\rho AM$ for some scalar $\rho > 0$,

then M is a Q-matrix if and only if the map $f_{\alpha, A, B}$ is

surjective. If we also assume that M (and therefore $f_{\alpha, A, B}$)

is weakly nondegenerate then the relationship between Q-matrices

and surjectivity can be sharpened as follows. Recall that on the

complementary cone $\overline{C}_K(\alpha)$ the function f is given by the linear

expression f_K. Define

$$P = \cup\{\overline{C}_K(\alpha): K \subset N, \det f_K' > 0\}$$
$$N = \cup\{\overline{C}_K(\alpha): K \subset N, \det f_K' < 0\}$$
$$Z = \cup\{\overline{C}_K(\alpha): K \subset N, \det f_K' = 0\} .$$

Then we have

Proposition 4.1. Let M be weakly nondegenerate and

$f_{\alpha, A, B}$ a PL map satisfying $(*)_M$. Then M is a Q-matrix if and

only if $f(P) = R^n$ or $f(N) = R^n$.

Proof. If $f(P) = R^n$ or $f(N) = R^n$, then f is surjective

and hence M is a Q-matrix.

Conversely, assume M is a Q-matrix. Then f is surjective so that

$$R(\alpha,\ A,\ B)\ =\ f(P)\ \cup\ f(N)\ \cup\ f(Z)\ =\ R^n\ .$$

Since
$$f(H(\alpha))\ =\ f(\cup\{\partial\overline{C}_K(\alpha)\colon K \subset N\})$$

$$=\ \cup\{\partial f(\overline{C}_K(\alpha))\colon K \subset N\}$$

and each $f(\overline{C}_K(\alpha))\ =\ <A_{\overline{K}},\ B_K>$ is closed, $f(H(\alpha))$ is closed and nowhere dense.

Suppose $f(P) \neq R^n$. If $R^n - f(H(\alpha)) \subset f(P)$, then

$$R^n\ =\ \overline{R^n\ -\ f(H(\alpha))}\ \subset\ \overline{f(P)}\ =\ f(P)\ ,$$

a contradiction. Hence

$$R^n\ -\ f(P)\ -\ f(H(\alpha))\ \neq\ \phi\ .$$

Since $f(Z) \subset f(H(\alpha))$,

$$f(N)\ -\ f(P)\ -\ f(H(\alpha))\ =\ R^n\ -\ f(P)\ -\ f(H(\alpha))\ \neq\ \phi\ .$$

Therefore $cd(f) < 0$ by Proposition 3.3.

Similarly, if $f(N) \neq R^n$, then $cd(f) > 0$. Hence $f(P) = R^n$ or $f(N) = R^n$. $\square$

This proposition provides a new and easy necessary condition for a weakly nondegenerate matrix M to be in Q. Since at least $n + 1$ polyhedral cones are required to cover R^n, M is in Q

only if at least $n + 1$ principal minors have the same nonzero sign.

It follows from Proposition 3.3 that if M is weakly nondegenerate and $cd(f_{\alpha,A,B}) \neq 0$ for some PL map $f_{\alpha,A,B}$ satisfying $(*)_M$, then $M \in Q$. In fact, M is then in the interior of Q, as we show in

Proposition 4.2. Let M be a weakly nondegenerate matrix and $f_{\alpha,A,B}$ any PL map satisfying $(*)_M$. If $cd(f_{\alpha,A,B}) \neq 0$, then $M \in$ int Q.

Proof. Assume $B = -\rho AM$, $\rho > 0$, and $cd(f_{\alpha,A,B}) \neq 0$. Let D be any bounded, open set containing 0. Then

$$cd(f_{\alpha,A,B}) = deg(f_{\alpha,A,B}, \; D, \; 0) \neq 0 \; .$$

There is an $\varepsilon_1 > 0$ such that

$$deg(g, \; D, \; 0) = deg(f_{\alpha,A,B}, \; D, \; 0)$$

for any continuous map $g: R^n \to R^n$ satisfying

$$\| g - f_{\alpha,A,B} \| = \sup\{\| g(x) - f_{\alpha,A,B}(x) \| : \| x \| = 1\} \leq \varepsilon_1$$

(see Ortega and Rheinboldt [29]). Since the set of weakly nondegenerate matrices is open in $R^{n \times n}$ (see Tamir [38]), there is an $\varepsilon_2 > 0$ such that $\| \hat{M} - M \| \leq \varepsilon_2$ implies $\hat{M}$ is weakly nondegenerate. Let $\varepsilon = \min\{\varepsilon_1, \varepsilon_2\}$. There is a $c \geq 1$ such that $\| \hat{M} - M \| < \varepsilon/c$ implies

$$\| f_{\alpha,A,\hat{B}} - f_{\alpha,A,B} \| \leq \epsilon \; ,$$

where $\hat{B} = -\rho A \hat{M}$. Then $\| \hat{M} - M \| < \epsilon / c$ implies

$$cd(f_{\alpha,A,\hat{B}}) = cd(f_{\alpha,A,B}) \neq 0$$

and hence $\hat{M} \in Q$. Therefore $M \in \text{int } Q$.   □

In view of Proposition 3.3, we have the following simple sufficiency test for a weakly nondegenerate matrix M to be in Q. Take any $f_{\alpha,A,B}$ satisfying $(*)_M$ (e.g., $f_{I,I,-M}$) and any $y \in R^n - f(H(\alpha))$. For each $K \subseteq N$ such that $\det M_{KK} \neq 0$, compute $y^K = L f_K^{-1}(y)$. If this unique point is in $\bar{C}_K(\alpha)$, it is in $f^{-1}(y)$. All points in $f^{-1}(y)$ are determined in this way. Now if

$$\sum_{\{K : y^K \in f^{-1}(y)\}} \text{sgn det } f_K' \neq 0 \; ,$$

then $M \in \text{int } Q$.

If $R(\alpha, A, B)$ is not all of R^n it will have a boundary and this boundary will consist of parts of images of $(n-1)$-dimensional facets of certain cones $\bar{C}_K(\alpha)$ (i.e., $\partial R(\alpha, A, B) \subseteq f(H(\alpha))$). Moreover, any such facet must be nonseparating (i.e., the determinants associated with the adjacent cones having this common facet may not have the same nonzero sign). To show this, we first prove

Lemma 4.3. Let $f_{\alpha,A,B}$ be any PL map and $\bar{C}_K(\alpha)$ and $\bar{C}_J(\alpha)$ any two adjacent cones with common facet

$\langle \eta_{K,1}, \ldots, \eta_{K,n-1} \rangle$. If this is a separating facet, then

$$G = \left\{ y \in R^n : y = \sum_{i=1}^{n-1} a_i \xi_{K,i}, \ a_i > 0 \right.$$
$$\left. \text{for } i = 1, \ldots, n-1 \right\} \subset \text{int } R(\alpha, A, B) .$$

Proof. Assume $\langle \eta_{K,1}, \ldots, \eta_{K,n-1} \rangle$ is a separating facet so that

$$\text{sgn det } \widetilde{f}_K = \text{sgn det } \widetilde{f}_J \neq 0 .$$

Then $\xi_{J,1}, \ldots, \xi_{J,n}$ are linearly independent, hence

$$\xi_{K,n} = \sum_{i=1}^{n} b_i \xi_{J,i}$$

for some $b_1, \ldots, b_n \in R$. It follows that

$$b_n = - \frac{\text{det } \widetilde{f}_K}{\text{det } \widetilde{f}_J} < 0 .$$

Let

$$y = \sum_{i=1}^{n-1} a_i \xi_{K,i} \in G$$

and define

$$\delta_1 = \min\{a_1, \ldots, a_{n-1}\} > 0$$

$$\delta_2 = \min\{\delta_1/2b_i : b_i > 0, \ i = 1, \ldots, n-1\} > 0 .$$

Since $\xi_{K,1}, \ldots, \xi_{K,n}$ are linearly independent, we can define a linear transformation $T: R^n \to R^n$ by

$$T\xi_{K,i} = I_i, \quad i = 1, \ldots, n .$$

Define

$$\varepsilon = \frac{1}{\|T\|_L} \min\{\delta_1/2, \delta_2\} > 0 .$$

Suppose $\|\hat{y} - y\| < \varepsilon$, where

$$\hat{y} = \sum_{i=1}^{n} \hat{a}_i \xi_{K,i} \in R^n, \quad \hat{a}_1, \ldots, \hat{a}_n \in R .$$

Then

$$|\hat{a}_i - a_i| < \min\{\delta_1/2, \delta_2\}$$

for $i = 1, \ldots, n$ (where $a_n = 0$). For $i = 1, \ldots, n - 1$,

$$\hat{a}_i > \delta_1/2 .$$

If $\hat{a}_n \geq 0$, then

$$\hat{y} \in f(\overline{C}_K(\alpha)) \subset R(\alpha, A, B) .$$

Otherwise $\hat{a}_n < 0$ and

$$\hat{a}_n = -|\hat{a}_n - a_n| > -\delta_2 .$$

We have

$$\hat{y} = \sum_{i=1}^{n-1} (\hat{a}_i + \hat{a}_n b_i) \xi_{J,i} + \hat{a}_n b_n \xi_{J,n} .$$

Since $b_n < 0$, $\hat{a}_n b_n > 0$. If $b_i \leq 0$ for $i \neq n$, then

$$\hat{a}_i + \hat{a}_n b_i \geq \hat{a}_i > 0$$

If $b_i > 0$ for $i \neq n$, then

$$\hat{a}_i + \hat{a}_n b_i > \frac{\delta_1}{2} - \delta_2 b_i \geq \frac{\delta_1}{2} - \frac{\delta_1}{2b_i} \cdot b_i = 0 .$$

Hence

$$\hat{y} \in f(\overline{C}_J(\alpha)) \subset R(\alpha, A, B) .$$

Therefore $y \in \text{int } R(\alpha, A, B)$.   □

Now we can prove

　　　Proposition 4.4. Let $f_{\alpha,A,B}$ be any PL map. If

$y \in \partial R(\alpha, A, B)$, then there is an $x \in f^{-1}(y)$ such that x is

on a nonseparating facet in $H(\alpha)$.

　　　Proof. Let $y \in \partial R(\alpha, A, B)$. If $y \in f(\overline{C}_K(\alpha))$ and

$\det f_K' = 0$, then the proposition is satisfied since $f(\overline{C}_K(\alpha)) =$

$f(\partial \overline{C}_K(\alpha))$. Hence we may assume that $\det f_K' \neq 0$ for all $K \subset N$

such that $y \in f(\overline{C}_K(\alpha))$. Define

$$Y = \{V \subset H(\alpha): y \notin f(V), \quad V = \langle n_{i(1)}, \ldots, n_{i(n-1)} \rangle,$$

$$n_{i(j)} = \alpha_{i(j)} \text{ or } -\alpha_{i(j)}, \, i(j) = 1, \ldots, n \text{ for all } j\} ,$$

$$\delta = \min\{\text{dist}(y, f(V)): V \in Y\} > 0 ,$$

and

$$B_\delta(y) = \{\hat{y} \in R^n: \|\hat{y} - y\| < \delta\} .$$

Let

$$\hat{H}(\alpha) = \cup \{<\eta_{i(1)}, \ldots, \eta_{i(n-2)}>: \eta_{i(j)} = \alpha_{i(j)}$$
$$\text{or} -\alpha_{i(j)}, \; i(j) = 1, \; \ldots, \; n \; \text{for all} \; j\}.$$

Then $\hat{H}(\alpha)$ is the union of the $(n-2)$-faces of the cones $\overline{C}_K(\alpha)$,
$K \subseteq N$. Since $f(\hat{H}(\alpha))$ is contained in a finite union of $(n-2)$-
dimensional subspaces of R^n, $B_\delta(y) - f(\hat{H}(\alpha))$ is connected.
Because $R(\alpha, A, B)$ is closed, $y \in R(\alpha, A, B)$ and hence there
is a $K \subseteq N$ such that $y \in f(\overline{C}_K(\alpha))$. By assumption,
det $f'_K \neq 0$. Then, as in the proof of Lemma 4.3, we may define the
nonsingular linear transformation $T: R^n \rightarrow R^n$ by

$$T\xi_{K,i} = I_i, \quad i = 1, \; \ldots, \; n \; .$$

There are $a_i \geq 0$, $i = 1, \; \ldots, \; n,$ such that

$$y = \sum_{i=1}^{n} a_i \xi_{K,i} \; .$$

For any $\hat{y} = \sum_{i=1}^{n} \hat{a}_i \xi_{K,i}$,

$$\|\hat{y} - y\| \leq \|T^{-1}\|_L \cdot [\sum_{i=1}^{n} (\hat{a}_i - a_i)^2]^{1/2} \; .$$

Clearly we can find $\hat{a}_i > 0$, $i = 1, \; \ldots, \; n,$ such that

$$\|\hat{y} - y\| < \delta \; .$$

Then

$$\hat{y} \in \text{int} \; f(\overline{C}_K(\alpha)) \subseteq \text{int} \; R(\alpha, A, B)$$

so that

$$\hat{y} \in \text{int}(B_\delta(y) \cap R(\alpha, A, B)) \; .$$

Since   int $f(\hat{H}(\alpha)) = \phi$,

$$[B_\delta(y) - f(\hat{H}(\alpha))] \cap R(\alpha, A, B) = [B_\delta(y) \cap R(\alpha, A, B)] - f(\hat{H}(\alpha)) \neq \phi \;.$$

But

$$[B_\delta(y) - f(\hat{H}(\alpha))] \cap R(\alpha, A, B)^c = B_\delta(y) \cap R(\alpha, A, B)^c \neq \phi$$

since $y \in \partial R(\alpha, A, B)$. Therefore

$$[B_\delta(y) - f(\hat{H}(\alpha))] \cap \partial R(\alpha, A, B) \neq \phi \;.$$

Let

$$z \in [B_\delta(y) - f(\hat{H}(\alpha))] \cap \partial R(\alpha, A, B) \;.$$

Since $z \in \partial R(\alpha, A, B)$,   there is a $w \in \partial \overline{C}_K(\alpha)$ for some $K \subseteq N$

such that $f(w) = z$. We may assume without loss of generality

that

$$w = \sum_{i=1}^{n-1} b_i n_{K,i}$$

for some $b_i \geq 0$, $i = 1, \ldots, n-1$. In fact, $b_i > 0$ for

$i = 1, \ldots, n-1$ since $z \notin f(\hat{H}(\alpha))$. Moreover, $z \in B_\delta(y)$ so

that $y \in f(\langle n_{K,1}, \ldots, n_{K,n-1} \rangle)$. The facet $\langle n_{K,1}, \ldots, n_{K,n-1} \rangle$

cannot be separating, since this would imply $z \in$ int $R(\alpha, A, B)$

by Lemma 4.3.   □

It can be shown that there is a finite process for determin-

ing whether or not a given vector is in int $R(\alpha, A, B)$. However,

this process is much simpler for generators (columns of (A, B))

and will be given in the sequel for such vectors. Clearly, if

M is a Q-matrix, then every column of (A, B) is in

int R(α, A, B), for any PL map $f_{\alpha, A, B}$ satisfying $(*)_M$.

Previous literature on linear complementarity has considered the

property that every column of (A, B) is interior to one of the

complementary cones generated by (A, B). The condition that every

column of (A, B) is in int R(α, A, B) is a weaker condition.

If a column is interior to some complementary cone then it is in

int R(α, A, B), but it may also be in int R(α, A, B) when it is

not interior to any complementary cone.

It is now useful to define a nondegeneracy condition stronger

than the usual notion (i.e., all principal minors of M are non-

zero; or, equivalently, all complementary cones have full

dimension). We say that a set of n distinct generators is

almost complementary if the set contains at least n - 1

complementary generators. We say that a matrix M is strongly

nondegenerate if any set of n almost complementary generators

from (I, -M) is linearly independent. (If a matrix M is

nondegenerate, any set of n complementary generators from

(I, -M) is linearly independent; hence strong nondegeneracy

clearly implies nondegeneracy.) Note that for n = 3, M is

strongly nondegenerate if and only if every set of n distinct

generators from (I, -M) is linearly independent.

The above discussion of the boundary of R(α, A, B) now

leads to a new conjecture: ∂R(α, A, B) $\neq \phi$ implies there is some

column of (A, B) contained in ∂R(α, A, B). Then M $\in$ Q if and only if every column of (A, B) is in int $R(\alpha, A, B)$ for some $f_{\alpha,A,B}$ satisfying $(*)_M$. It will be shown later that this condition is necessary and sufficient for any strongly nondegenerate M and $n = 2$ or 3. (The case $n = 1$ is a trivial problem.) It is not known in general.

It still remains to discuss a way of checking whether or not a generator is in int $R(\alpha, A, B)$. These conditions will amount to determining whether or not a lower dimensional matrix is in Q. This will be more easily developed after several other results are demonstrated.

We now state without proof

Proposition 4.5.

A. For $n = 1$, M is a Q-matrix if and only if $m_{11} > 0$, which is true if and only if $cd(f_{\alpha,A,B}) = \pm 1$ (for any PL map f satisfying $(*)_M$).

B. Every Q-matrix in R^2 is weakly nondegenerate (although not necessarily nondegenerate) and, in R^2, $M \in Q$ if and only if $cd(f) = \pm 1$ for any $f_{\alpha,A,B}$ satisfying $(*)_M$.   []

These facts will be of use in proving a main theorem of this section. This theorem also reveals the process of checking (via a projection) whether or not a generator is in int $R(\alpha, A, B)$.

<u>Theorem 4.6.</u>   Let $f_{\alpha,A,B}$ be a PL map satisfying $(*)_M$,

where   M   is a <u>strongly nondegenerate</u> 3×3 matrix.   Then $M \in Q$

$\implies cd(f_{\alpha,A,B}) \neq 0$.

<u>Proof.</u>   First, let us observe some geometrical facts.

A.   Let $\xi_1, \xi_2, \xi_3, \xi_1', \xi_2', \xi_3'$ denote the six distinct

generators of the image cones (i.e., the columns of   A   and   B)

where $\{\xi_i, \xi_i'\} = \{A_1, B_1\}$, $\{A_2, B_2\}$ or $\{A_3, B_3\}$ for   i = 1,

2, 3.   For any $z \in R^3$ let $\hat{z}$ denote $z/\|z\|$,   a vector on the

unit sphere S^2.   Each closed complementary cone, say

$\langle \xi_1, \xi_2, \xi_3 \rangle$,   intersects with S^2 to form a closed spherical

triangle denoted $>\xi_1, \xi_2, \xi_3<$.   Let   p   be the stereographic pro-

jection of $S^2 - \{-\hat{\xi}_3\}$ onto the plane R^2,   where we take $\hat{\xi}_3$ to

be the north pole $(p\hat{\xi}_3 = 0)$ and $-\hat{\xi}_3$ to be the south pole

$(p(-\hat{\xi}_3) = \infty)$.   See   Figure 4.1.   Now a PL map $\tilde{f}: R^2 \to R^2$ is

induced as follows.

For $x \in R^2$,   let us define

$$\tilde{f}(x) = \tilde{f}_{\tilde{\alpha},\tilde{A},\tilde{B}}(x) - \sum_{\{i:x_i > 0\}} x_i p\hat{\xi}_i - \sum_{\{i:x_i < 0\}} x_i p\hat{\xi}_i'$$

where $\tilde{f}_{\tilde{\alpha},\tilde{A},\tilde{B}}: R^2 \to R^2$ is the PL map determined by

$\tilde{\alpha} = I$, $\tilde{A} = [p\hat{\xi}_1, p\hat{\xi}_2]$,   and $\tilde{B} = [p\hat{\xi}_1', p\hat{\xi}_2']$.   (The strong nonde-

generacy assumption guarantees that $\{\hat{\xi}_1, \hat{\xi}_2, \hat{\xi}_3, \hat{\xi}_1', \hat{\xi}_2', \hat{\xi}_3'\} \subset$

$S^2 - \{-\hat{\xi}_3\}$.)   We observe that the nondegeneracy of f   is inherited

by $\tilde{f}$.   That is, suppose some cone generated by $(\tilde{A}, \tilde{B})$,   say

$\langle p\hat{\xi}_1, p\hat{\xi}_2 \rangle$,   has dimension less than 2 (where $\langle p\hat{\xi}_1, p\hat{\xi}_2 \rangle$ is

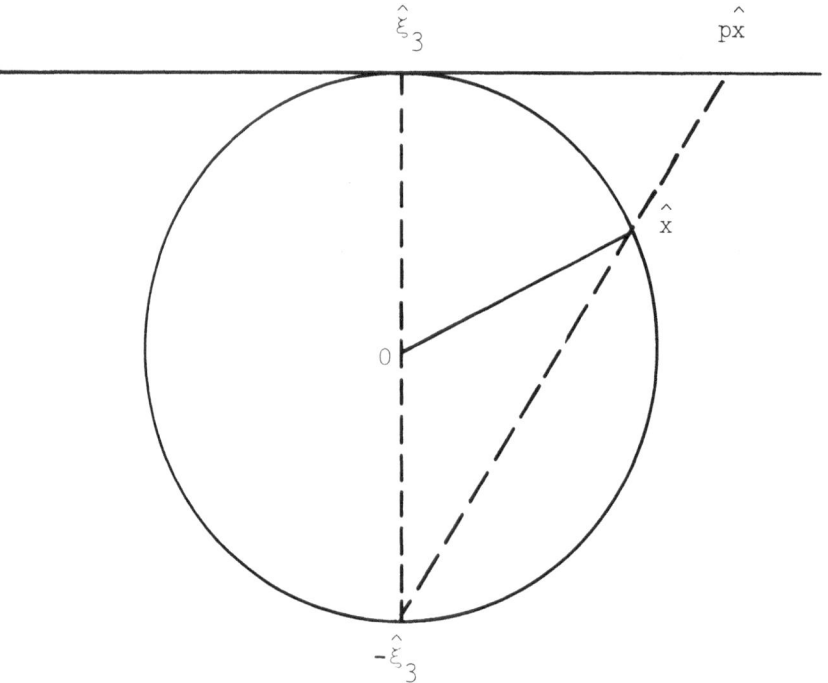

Figure 4.1

The Projection, $p: S^2 - \{-\hat{\xi}_3\} \to R^2$, $p(\hat{\xi}_3) = (0, 0)$.

If $\hat{\xi}_3 = A_3 = I_3$, then $p\hat{x} = \dfrac{2}{1 + \hat{x}_3} \begin{bmatrix} \hat{x}_1 \\ \hat{x}_2 \end{bmatrix}$, for any $\hat{x} \in S^2 - \{-I_3\}$.

understood to be a cone in R^2 relative to the origin $(0, 0)$

$= p\hat{\xi}_3)$. Then $p\hat{\xi}_1$, $p\hat{\xi}_2$, $p\hat{\xi}_3$ all lie on the same ray through the

origin of R^2 and hence $\hat{\xi}_1$, $\hat{\xi}_2$, $\hat{\xi}_3$ lie on the same great circle

of S^2 (through the north pole), which is impossible since

$\langle\xi_1, \xi_2', \xi_3\rangle$ has dimension 3.

Now consider $\overline{C}_\phi(\tilde{\alpha}) = \langle I_1, I_2\rangle$, and any adjacent cone, say

$\overline{C}_2(\tilde{\alpha}) = \langle I_1, -I_2\rangle$. Clearly, int $\tilde{f}(\overline{C}_\phi(\tilde{\alpha}))$ and int $\tilde{f}(\overline{C}_2(\tilde{\alpha}))$

intersect if and only if det $\tilde{f}_\phi'$ and det $\tilde{f}_2'$ have opposite

nonzero signs. But int $\tilde{f}(\overline{C}_\phi(\tilde{\alpha})) = \mathrm{int}\langle p\hat{\xi}_1, p\hat{\xi}_2\rangle$ and

int $\tilde{f}(\overline{C}_2(\tilde{\alpha})) = \mathrm{int}\langle p\hat{\xi}_1, p\hat{\xi}_2'\rangle$ intersect if and only if

$\mathrm{int}\langle\xi_1, \xi_2, \xi_3\rangle$ and $\mathrm{int}\langle\xi_1, \xi_2', \xi_3\rangle$ intersect (because the

stereographic projection p is one to one). Hence these latter

cones intersect if and only if det $f_{K(\phi)}'$ and det $f_{K(\{2\})}'$ have

opposite nonzero signs, where $K(J) \subset \{1, 2, 3\}$ is defined for

any $J \subset \{1, 2\}$ by

$$i \in K(J) \text{ if and only if } \begin{cases} i = 1 \text{ or } 2, \ i \in J, \text{ and } \xi_i = A_i; \text{ or} \\ i = 1 \text{ or } 2, \ i \notin J, \text{ and } \xi_i = B_i; \text{ or} \\ i = 3 \text{ and } \xi_i = B_i \end{cases}$$

so that $f(\overline{C}_{K(\phi)}(\alpha)) = \langle\xi_1,\xi_2,\xi_3\rangle$, $f(\overline{C}_{K(\{2\})}(\alpha)) = \langle\xi_1,\xi_2',\xi_3\rangle$,

etc. From this it is readily deduced that

(i) if det $\tilde{f}_\phi'$ and det $f_{K(\phi)}'$ have the same sign, then

for all $J \subset \{1, 2\}$, sgn det $\tilde{f}_J' = $ sgn det $f_{K(J)}'$

(ii) if det $\tilde{f}_\phi'$ and det f_ϕ' have opposite signs, then for

all $J \subset \{1, 2\}$, sgn det $\tilde{f}_J' = -$sgn det $f_{K(J)}'$.

B. Let $c(A, B) = >\xi_1, \xi_2< \cup >\xi_2, \xi_1'< \cup >\xi_1', \xi_2'< \cup >\xi_2', \xi_1'<$

where, for example, $>\xi_1, \xi_2<$ denotes the intersection of the

facet $<\xi_1, \xi_2>$ with S^2, etc. Assuming (A, B) is nondegen-

erate, the complement in S^2 of $c(A, B)$ consists of 2 or 3

components (Figure 4.2(a) or 4.2(b), respectively; see Kelly and

Watson [13]). We define

$$C(\xi_3) = \{\hat{x} \in S^2 \colon \hat{x} \notin \cup C<\xi_3>\} .$$

($C<\xi_3>$ is the set of complementary extensions of ξ_3 and

$\cup C<\xi_3> = \cup\{G \colon G \in C<\xi_3>\}.$) Observe that

$$S^2 - R(\alpha, A, B) = C(\xi_3) \cap C(\xi_3') .$$

C. If $\xi_3 \in \text{int}(\cup C<\xi_3>)$, then from Figures 4.1 and 4.2, if

we project S^2 onto R^2 using $\hat{\xi}_3$ as north pole, there are only

five configurations possible, namely Figures 4.3(a)-(e). The

shaded region is the set $pC(\xi_3)$, which is equal to

$$pC(\xi_3) = \{x \in R^2 \colon \lambda x \notin pc(A, B), \text{ all } \lambda \geq 1\} .$$

In Figures 4.3(a)-(d), at least three generators (among

$p\hat{\xi}_1, p\hat{\xi}_2, p\hat{\xi}_1', p\hat{\xi}_2'$) are in the boundary of $pC(\xi_3)$, while in

Figure 4.3(e) only two generators are in the boundary of $pC(\xi_3)$.

D. If $\xi_3 \notin \text{int}(\cup C<\xi_3>)$, Figures 4.4(a)-(c) show all the

configurations possible. The shaded regions are points in

$pC(\xi_3)$. In Figures 4.4(a) and (b), the convex cone

$<p\hat{\xi}_1, p\hat{\xi}_2>$ contains the projection of $\cup C<\xi_3> \cap S^2$, whereas in

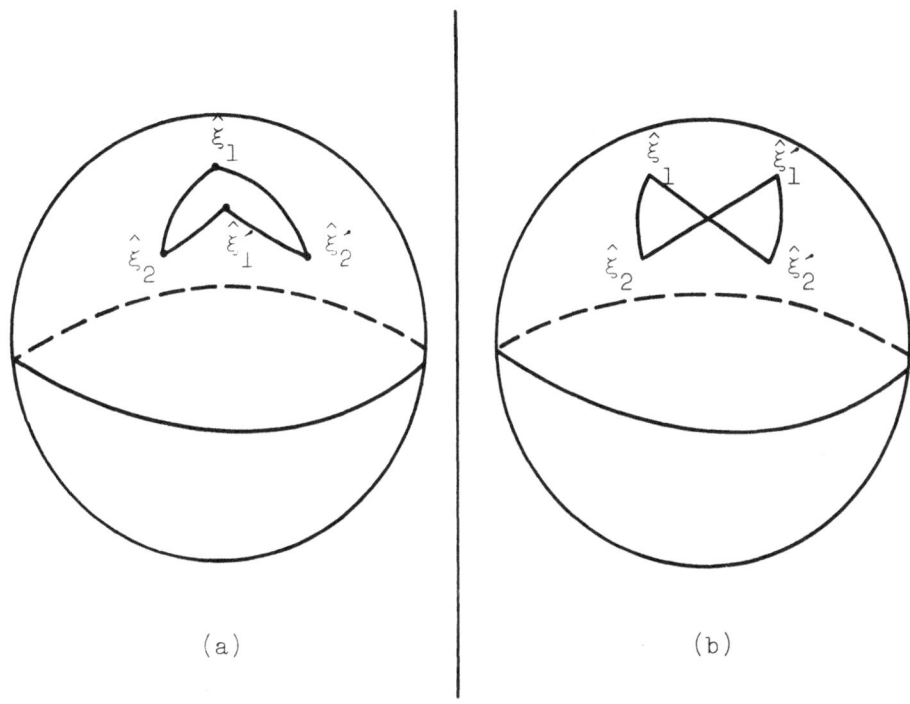

Figure 4.2

$c(\Lambda, B)$ divides S^2 into two or three components

In all cases, $p\hat{\xi}_3 = (0,0)$.

Figure   4.3

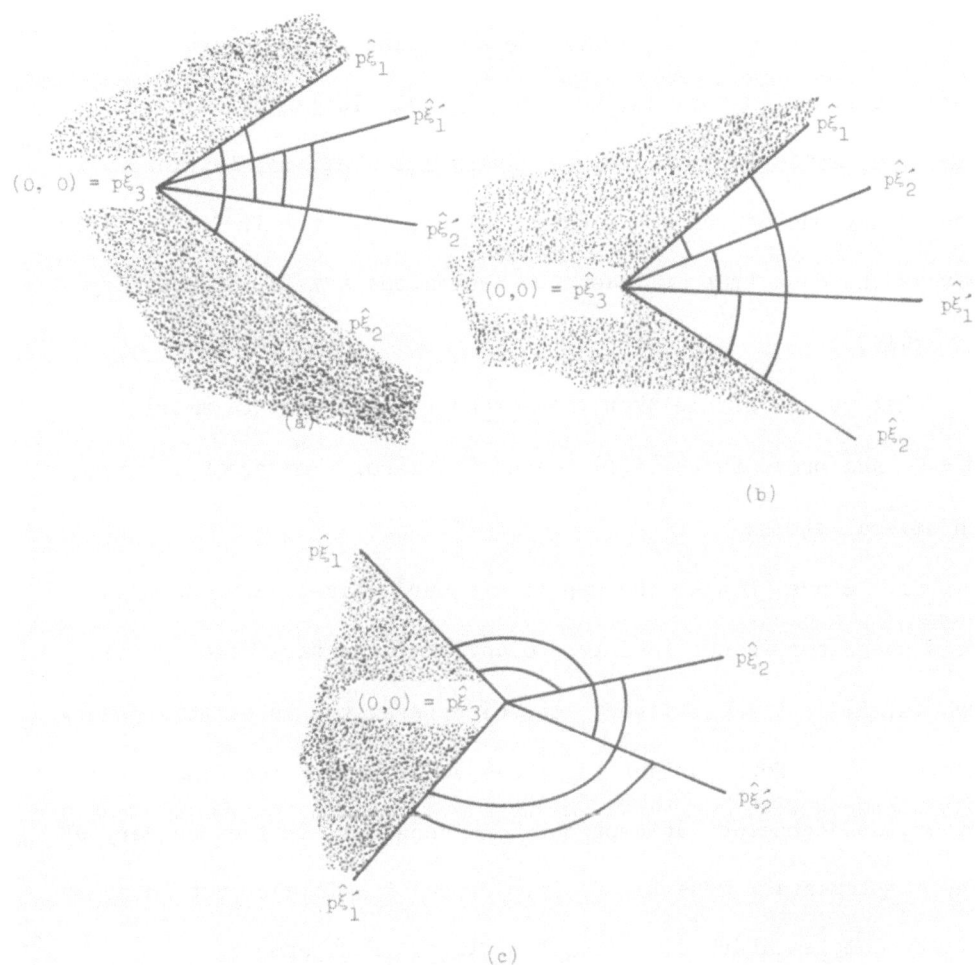

Figure 4.4

Figure 4.4(c), the convex cone $\langle \hat{p\xi}_1, \hat{p\xi}_1' \rangle$ need not. Also, all

three configurations have at least two generators in the boundary

of $p\mathcal{C}(\xi_3)$, namely $\hat{p\xi}_1$ and $\hat{p\xi}_2$ in (a) and (b), and $\hat{p\xi}_1$ and

$\hat{p\xi}_1'$ in (c).

E. Overall, it is always the case that at least two

generators are in the boundary of $p\mathcal{C}(\xi_3)$. It can be observed

that this will remain true more generally. That is, letting ν

denote any one of the six generators ξ_i, ξ_i', $i = 1, 2, 3,$ it is

always the case that at least two generators are in the boundary

of $p\mathcal{C}(\nu)$.

Let us now assume that the <u>strongly</u> nondegenerate matrix

$M \in Q$ and prove that $cd(f)$ cannot be zero. The proof is given

in several steps.

1. Since $M \in Q$ the map f is surjective. Suppose that

some generator is not interior to any complementary cone spanned

by (A, B). We may, without loss of generality, label this gener-

ator as ξ_3. Now consider the above described projection. By the

strong nondegeneracy assumption, ξ_3 cannot be on the boundary of

any complementary cone in $C\langle \xi_3' \rangle$. Since ξ_3 is also not interior

to any complementary cone, $\xi_3 \notin \cup C\langle \xi_3' \rangle$. Furthermore,

$$\xi_3 \in int\ R(\alpha, A, B) = int\big((\cup C\langle \xi_3 \rangle) \cup (\cup C\langle \xi_3' \rangle) \big)$$

so that $\xi_3 \in int(\cup C\langle \xi_3 \rangle)$ since $\cup C\langle \xi_3' \rangle$ is closed. This means

that the four complementary planar cones $\langle \hat{p\xi}_1, \hat{p\xi}_2 \rangle$, $\langle \hat{p\xi}_1, \hat{p\xi}_2' \rangle$,

$\langle \hat{p\xi}_1', \hat{p\xi}_2 \rangle$, and $\langle \hat{p\xi}_1', \hat{p\xi}_2' \rangle$ must cover R^2 (see Figure 4.3) and hence, by the Proposition 4.5, $cd(\tilde{f}) = +1$ or -1.   But for any

nonzero $y \in R^2$ with sufficiently small norm and lying in the

boundary of no planar cone, the planar cones that cover y

correspond one-to-one with the images of the spherical triangles

which cover y and these obviously correspond one-to-one with the

spherical triangles that cover $p^{-1}y$. Since these correspond to

the only cones containing $p^{-1}y$, and since either sgn det $\tilde{f}'_J$

$= $ sgn det $f'_{K(J)}$ for all $J \subseteq \{1,\ 2\}$ or sgn det $\tilde{f}'_J =$

$-$sgn det $f'_{K(J)}$ for all $J \subseteq \{1,\ 2\}$, we conclude that $cd(f) = +1$

or -1.

2.   Otherwise each generator is <u>interior</u> to some

complementary cone.   Consider three exhaustive and mutually

exclusive cases:

(i)   Either of Figures 4.3(a)-(d) occurs.

(ii)   Either of Figures 4.4(a)-(c) occurs.

(iii)   Figure 4.3(e) occurs.

3.   For case (i), the projections of at least three genera-

tors (among $\hat{\xi}_1$, $\hat{\xi}_2$, $\hat{\xi}'_1$, $\hat{\xi}'_2$)   are in the boundary of $pC(\xi_3)$.   In

accordance with the observation in part (E) above, the projections

of at least two of these generators are in the boundary of

$pC(\xi'_3)$. Hence there must be a generator in the boundary of

$pC(\xi_3)$ and also in the boundary of $pC(\xi'_3)$.   Thus, this generator

cannot be interior to some complementary extension of ξ_3 or of

ξ'_3, i.e., this generator cannot be interior to any complementary

cone, a contradiction.

4. Now, consider case (ii). Since $\xi_3^{'}$ is interior to a

complementary cone, $p\hat{\xi}_3^{'} \in <p\hat{\xi}_1, p\hat{\xi}_2>$ in Figures 4.4(a) and (b)

and $p\hat{\xi}_3^{'} \notin <p\hat{\xi}_1, p\hat{\xi}_1^{'}>$ in Figure 4.4(c). But if Figure 4.4(a) or

(b) were to occur, then we could not have ξ_3 interior to a

complementary cone. Thus only Figure 4.4(c) is possible. In this

case, $p\hat{\xi}_3^{'} \notin <p\hat{\xi}_2, p\hat{\xi}_2^{'}>$ since $\xi_3 \in {}^{\cup}C<\xi_3^{'}>$. Hence, without loss

of generality, take $p\hat{\xi}_3^{'} \in <p\hat{\xi}_1, p\hat{\xi}_2>$. Then

$\xi_1^{'} \notin <\xi_1, \xi_2, \xi_3^{'}> {}^{\cup} <\xi_1, \xi_2^{'}, \xi_3^{'}>$. But this implies that $\xi_1^{'}$ is

not contained in the interior of any complementary extension of

$\xi_3^{'}$. Hence, since $p\hat{\xi}_1^{'}$ is in the boundary of $p^C(\xi_3)$ (see part

(D) above), $\xi_1^{'}$ is not interior to any complementary cone, a

contradiction.

5. Finally, consider case (iii). Clearly, $-\xi_3^{'}$ is not

contained in any complementary extension of $\xi_3^{'}$. Since f is

surjective, $-\xi_3^{'}$ must be contained in at least one complementary

extension of ξ_3. By the strong nondegeneracy assumption, $-\xi_3^{'}$

cannot be on the boundary of any complementary extension of

ξ_3. If $-\xi_3^{'}$ is in one or three complementary extensions of

ξ_3, then cd(f) $\neq$ 0. Otherwise $-\xi_3^{'}$ is contained in two

complementary extensions of ξ_3. In fact, $p(-\hat{\xi}_3)$ must be inside

the dotted region determined by the points a, $p\hat{\xi}_1$ and $p\hat{\xi}_2$ in

Figure 4.3(e). Thus $<\xi_1^{'}, \xi_2, \xi_3>$ and $<\xi_1, \xi_2^{'}, \xi_3>$ are the

only two complementary cones containing $-\xi_3^{'}$. Since cd($\tilde{f}$) $\neq$ 0

($\tilde{f}$ is surjective in this case), it follows from the geometry of

Figure 4.3(e) that on the cone $<p\hat{\xi}_1^{'}, p\hat{\xi}_2>$ in R^2 sgn det $\tilde{f}^{'}$ is nonzero and is equal to $\operatorname{sgn} \det \tilde{f}$ on the cone $\langle p\hat{\xi}_1, p\hat{\xi}_2' \rangle$.

Hence we have that $\operatorname{sgn} \det f$ on $\langle \xi_1', \xi_2, \xi_3 \rangle$ is equal to

$\operatorname{sgn} \det f$ on $\langle \xi_1, \xi_2', \xi_3 \rangle$. Thus $cd(f) = \pm 2$. $\square$

Consider the case where $f_{\alpha,A,B} = f_{I,I,-M}$. We have just

shown that if a strongly nondegenerate $M \in R^{3\times 3}$ is in Q then

$cd(f_{I,I,-M}) \neq 0$. A slight perturbation of any such $M \in Q$, say

to $\hat{M}$, will be such that $\hat{M}$ is also strongly nondegenerate and,

by the properties of degree (cf. the proof of Proposition 4.2),

$cd(f_{I,I,-\hat{M}}) \neq 0$. Hence by Proposition 4.2, $\hat{M} \in \text{int } Q$. Thus the

known result that the class of strongly nondegenerate 3×3 Q-

matrices is open follows immediately. A natural question prompted

by Theorem 4.6 is whether or not it is true in general that a

strongly nondegenerate matrix $M \in Q$ if and only if $cd(f) \neq 0$.

(The if part follows from Proposition 4.2. Hence the question is

whether or not M strongly nondegenerate and $M \in Q$ imply

$cd(f) \neq 0$.) Closely related to this, Howe [10] asks if for a

weakly nondegenerate M, $M \in Q$ if and only if $cd(f) \neq 0$. The

example of Kelly and Watson [13] shows that this is not so. Let

$$
M(\varepsilon) = \begin{bmatrix}
21 & 25 & -27 & -36 - \varepsilon \\
7 & 3 & -9 & 36 + \varepsilon \\
12 & 12 & -20 & 0 \\
4 & 4 & -4 & -8
\end{bmatrix}.
$$

Kelly and Watson show that this (nondegenerate) matrix $M(0) \in Q$,

but that $M(\varepsilon) \notin Q$ for $0 < \varepsilon < 1$. (This matrix is not strongly

nondegenerate since I_1, I_2, I_4, $-M_4$ are linearly dependent.)

Hence $cd(f) = 0$ for any f satisfying $(*)_{M(0)}$. For if $cd(f)$

$\neq 0$, then any slight perturbation of $M(0)$ to $\hat{M}$ will produce a

new PL map $\hat{f}$ satisfying $(*)_{\hat{M}}$ and such that $cd(\hat{f}) = cd(f) \neq 0$,

which would imply that any slight perturbation of $M(0)$ is also

in Q.

In fact, using the techniques of Kelly and Watson (which are

not easy to describe here), it may be shown that $M(0)$ may also

be perturbed into the int Q. Hence the conjecture is still false

if we restrict M to only the interior of Q (i.e., it is false

that M nondegenerate and $M \in int \ Q$ imply $cd(f) \neq 0$).

Regarding Theorem 4.6, we conjecture that the theorem is true

if "strongly" is replaced by "weakly," whereas the theorem is

false if "3 × 3" is replaced by "4 × 4."

We now consider the conjecture that a strongly nondegenerate

matrix $M \in R^{n \times n}$ is a Q-matrix if and only if all columns of

(A, B) are in int $R(\alpha, A, B)$, which means in particular that

for each column, either it is interior to at least one complemen-

tary cone (spanned by (A, B)), or that the stereographic

projection, with that column as the north pole, produces a Q-

matrix of dimension $n - 1$. For example, suppose we want to know

whether some column, say ξ_j, is in int $R(\alpha, A, B)$. If it is

interior to some complementary cone (note that $\xi_j \quad \in$

$\text{int} < \xi_{j_1}, \ldots, \xi_{j_n} >$ if and only if $[\xi_{j_1}, \ldots, \xi_{j_n}]^{-1}\xi_j > 0)$ then certainly it is in int $R(\alpha, A, B)$. If it is not interior to any cone, make a stereographic projection onto R^{n-1} with $\hat{\xi}_j$ as the north pole and $-\hat{\xi}_j$ as the south pole. Then ξ_j is in int $R(\alpha, A, B)$ if and only if the $(n-1) \times (n-1)$ matrix $\tilde{M} \in Q$, where

$$\tilde{M} = -\tilde{A}^{-1}\tilde{B} ,$$

$\tilde{A}_i = p\hat{A}_i$, $\tilde{B}_i = p\hat{B}_i$, $1 \leq i \leq j - 1$, and $\tilde{A}_i = p\hat{A}_{i+1}$, $\tilde{B}_i = p\hat{B}_{i+1}$, $j \leq i \leq n-1$. For example it can be verified that if $f_{\alpha, A, B} = f_{I, I, -M}$, and I_n is not interior to any complementary cone, then I_n is in int $R(I, I, -M)$ if and only if the principal submatrix M_{KK}, $K = \{1, \ldots, n-1\}$, is a Q-matrix in

$$R^{(n-1) \times (n-1)}. \quad (\text{Here,} \quad p\hat{x} = \frac{2}{1 + \hat{x}_n}\begin{bmatrix} \hat{x}_1 \\ \vdots \\ \hat{x}_{n-1} \end{bmatrix}, \quad \text{for all} \quad \hat{x} \in S^{n-1},$$

$\hat{x} \neq -I_n)$.

The idea of relating a Q-matrix M to lower dimensional Q-matrices was initially introduced and studied by Cottle [1]. He defined "completely Q-matrices" to be those for which all principal submatrices are also in Q. In a quite different approach, we have indicated (via the above construction) that all Q-matrices are affiliated (via projection) with lower dimensional Q-matrices. We conclude this section with

Theorem 4.7. Let M be a strongly nondegenerate 3×3 matrix and $f_{\alpha, A, B}$ any PL map satisfying $(*)_M$. Then $M \in Q$ if and only if every column of (A, B) is in int $R(\alpha, A, B)$.

Proof. The necessity is immediate. To prove sufficiency, we assume that every column of (A, B) is in int $R(\alpha, A, B)$ and proceed in a manner similar to the proof of Theorem 4.6. If there is a generator which is not interior to any complementary cone, then (as shown in part (1) of the proof of Theorem 4.6) $cd(f) \neq 0$ and so $M \in Q$ by Proposition 4.2.

Otherwise each generator is interior to some complementary cone and we may conclude, as in parts (3) and (4) of the proof of Theorem 4.6, that Figure 4.3(e) must arise. If $\xi_3^{'} \in \cup C<\xi_3>$, then it is clear from Figure 4.3(e) that $\cup C<\xi_3^{'}> \subset \cup C<\xi_3>$. But then $\xi_1^{'}$ and $\xi_2^{'}$ cannot be in the interior of any complementary cone, a contradiction. Therefore $\xi_3^{'} \notin \cup C<\xi_3>$. But then $\xi_3^{'}$ cannot be in the interior of any complementary cone, a contradiction. Hence some generator is not interior to any complementary cone and so $M \in Q$.   □

The above theorem is trivial for 2×2 matrices. In fact, if $f_{\alpha, A, B}$ is any PL map from R^2 into R^2 for which every column of (A, B) is in int $R(\alpha, A, B)$, then f is surjective. For suppose there is a $q \in \partial R(\alpha, A, B)$. Then $q \in f(H(\alpha))$. If $q \neq 0$, then $\hat{q} = \hat{\xi}$ for some generator ξ, hence $q \in$ int $R(\alpha, A, B)$, a contradiction. Thus 0 is the only possible element of $\partial R(\alpha, A, B)$. But $\partial R(\alpha, A, B) = \{0\}$ implies

$R(\alpha, A, B) = \{0\}$ and int $R(\alpha, A, \cdot B) = \phi$, a contradiction.

Therefore $\partial R(\alpha, A, B) = \phi$ and hence $R(\alpha, A, B) = R^2$. It is not

known whether the theorem extends to $n > 3$.

REFERENCES

[1] Cottle, R. W., "Completely Q-Matrices," <u>Math. Progr.</u> 19
     (1980), 347-51.

[2] Cottle, R. W., and G. B. Dantzig, "Complementary Pivot
     Theory of Mathematical Programming," <u>Linear Algebra and Its</u>
     <u>Appls.</u> 1 (1968), 103-25.

[3] Cottle, R. W., G. J. Habetler, and C. E. Lemke, "Quadratic
     Forms Semi-Definite over Convex Cones," <u>Proc. of the</u>
     <u>International Symposium on Math. Progr.</u> Princeton (1967),
     551-65.

[4] Cottle, R. W., and R. von Randow, "On Q-matrices, Centroids,
     and Simplotopes," Stanford Tech. Rep. 79-10 (1979).

[5] Doverspike, R. D., and C. E. Lemke, "A Partial
     Characterization of a Class of Matrices Defined by Solutions
     to the Linear Complementarity Problem," Rensselaer
     Polytechnic Inst. Tech. Rep. (1979).

[6] Eaves, B. C., "The Linear Complementarity Problem," Management Science 17 (1971), 612-34.

[7] Eaves, B. C., and H. Scarf, "The Solution of Systems of Piecewise Linear Equations," Math. of O.R. 1 (1976), 1-27.

[8] Garcia, C. B., F. J. Gould, and T. R. Turnbull, "A PL Homotopy Method for the Linear Complementarity Problem," to be published in Proceedings of the International Congress on Mathematical Programming (ed. Milton Kelmanson), North-Holland.

[9] Heyden, L. van der, "A Variable Dimension Algorithm for the Linear Complementarity Problem," Math. Progr. 19 (1980), 328-46.

[10] Howe, R., "Linear Complementarity and the Degree of Mappings," Cowles Foundation Discussion Paper No. 542 (1980).

[11] Kaneko, I., "The Number of Solutions of a Class of Linear Complementarity Problems," Math. Progr. 17 (1979), 104-05.

[12] Karamardian, S. "The Complementarity Problem," Math. Progr. 2 (1972), 107-29.

[13] Kelly, L. M., and Watson, L. T., "Q-matrices and Spherical Geometry," Linear Algebra and Its Appls. 25 (1979), 175-89.

[14] Kojima, M., H. Nishino and T. Sekine, "An Extension of Lemke's Method to the Piecewise Linear Complementarity Problem," SIAM J. Appl. Math. 31 (1976), 600-13.

[15] Kojima, M., and R. Saigal, "On the Number of Solutions to a
      Class of Linear Complementarity Problems," Math. Progr. 17
      (1979), 136-39.

[16] Kojima, M., and R. Saigal, "On the Number of Solutions to a
      Class of Linear Complementarity Problems," Math. Progr. 21
      (1981), 190-203.

[17] Kostreva, M. M., "Direct Algorithms for Complementarity
      Problems," Ph.D. Dissertation, Rensselaer Polytechnic
      Institute (1976).

[18] Lemke, C. E., "On Complementary Pivot Theory," Math. of
      Decision Sciences, eds. G. B. Dantzig and A. F. Veinott,
      Jr., AMS-Providence (1968), 95-114.

[19] Lemke, C. E., "Recent Results on Complementarity Problems,"
      Nonlinear Programming, eds. O. L. Mangasarian and K. Ritter,
      Academic Press, New York (1970), 349-84.

[20] Lemke, C. E., and J. T. Howson, Jr., "Equilibrium Points of
      Bimatrix Games," SIAM Review 12 (1964), 413-23.

[21] Mangasarian, O. L., "Equivalence of the Complementarity
      Problem to a System of Nonlinear Equations," Univ. of
      Wisconsin Tech. Rep. No. 227 (1974).

[22] Mangasarian, O. L., "Linear Complementarity Problems
      Solvable by a Single Linear Program," Math. Progr. 10
      (1976), 263-70.

[23] Megiddo, N., and M. Kojima, "On the Existence and Uniqueness
      of Solutions in Nonlinear Complementarity Theory," Math.

Progr. 12 (1977), 110-130.

[24] Murty, K. G., "On the Number of Solutions to the Complementarity Problem and Spanning Properties of Complementary Cones," _Linear Algebra and Its Appls._ 5 (1972), 65-108.

[25] Murty, K. G., "Note on a Bard-type Algorithm for Solving the Complementarity Problem," _Opsearch_ 11 (1974), 123-130.

[26] Murty, K. G., "Some Results on Linear Complementarity Problems Associated with P-Matrices," Tech. Rep. No. 77-10, IOE Dept., Univ. of Michigan (1977).

[27] Murty, K. G., "On the Linear Complementarity Problem," _Proc. of the Third Symposium on Operations Research_, eds. W. Oettli and F. Steffens, Verlagrgruppe, Athenaum/Hain (1978), 425-439.

[28] Murty, K. G., "Computational Complexity of Complementary Pivot Methods," _Math. Programming Study_ 7 (1978), 61-73.

[29] Ortega, J. M., and W. C. Rheinboldt, _Iterative Solution of Nonlinear Equations in Several Variables_, Academic Press, New York (1970).

[30] Panne, C. van de, "A Complementary Variant of Lemke's Method for the Linear Complementarity Problem," _Math. Progr._ 7 (1974), 283-310.

[31] Pang, J. S., "On Q-matrices," _Math. Progr._ 17 (1979), 243-47.

[32] Pang, J. S., I. Kaneko and W. P. Hallman, "On the Solution

of Some (Parametric) Linear Complementarity Problems with

Applications to Portfolio Selection, Structural Engineering,

and Actuarial Graduation," Math. Progr. 16 (1979), 325-47.

[33] Saigal, R., "A Note on a Special Linear Complementarity

Problem," Opsearch 7 (1970), 175-183.

[34] Saigal, R., "A Characterization of the Constant Parity

Property of the Number of Solutions to the Linear Complemen-

tarity Problem," SIAM J. Appl. Math. 23 (1972), 40-45.

[35] Saigal, R., "On the Class of Complementary Cones and Lemke's

Algorithm," SIAM J. on Appl. Math. 23 (1972), 46-60.

[36] Saigal, R., and C. P. Simon, "Generic Properties of the

Complementarity Problem," Math. Progr. 3 (1973), 324-335.

[37] Shapley, L. S., "A Note on the Lemke-Howson Algorithm," Math

Progr. Study 1 (1974), 175-89.

[38] Tamir, A., "The Complementarity Problem of Mathematical

Programming," Ph.D. Dissertation, Case Western Reserve

University (1973).

[39] Todd, M. J., "A Generalized Complementary Pivoting

Algorithm," Math. Progr. 6 (1974), 243-63.

[40] Todd, M. J., "Orientation in Complementary Pivot

Algorithms," Math. of O.R. 1 (1976), 54-66.

[41] Watson, L. T., "A Variational Approach to the Linear

Complementarity Problem," Ph.D. Dissertation, Department of

Mathematics, University of Michigan (1974).

[42] Watson, L. T., "Some Perturbation Theorems for Q-Matrices," SIAM J. Appl. Math. 31 (1976), 379-384.

[43] Watson, L. T., "An Algorithm for the Linear Complementarity Problem," Intern. J. Computer Math. 6 (1978), 319-325.

A NOTE ON STEPSIZE CONTROL FOR NUMERICAL CURVE FOLLOWING

K. Georg

Institut für Angewandte Mathematik

Wegelerstr. 6, D-5300 Bonn

ABSTRACT

We sketch a simple strategy to monitor the stepsize in predictor-corrector methods for following curves which are implicitly defined by $H(x)=0$, where $H: \mathbb{R}^{N+1} \to \mathbb{R}^N$ is smooth. The control is based on elementary asymptotic error considerations and has been found to be quite successful.

(1) INTRODUCTION

We consider a sufficiently smooth map $H: \mathbb{R}^{N+1} \to \mathbb{R}^N$ such that zero is a regular value, and are interested in predictor-corrector methods in the sense of Haselgrove [7], which follow a solution curve $c(s)$ in $H^{-1}(0)$.

Partially supported by Deutsche Forschungsgemeinschaft, SFB 72 at Bonn, and by U.S.Air Force WPAFB under contract number FY 1456-81-00870.

Without loss of generality, we assume here that $c(s)$ is parametrized according to arc length.

Let us introduce the following definitions. If A is an $(N,N+1)$-matrix with maximal rank N, we define a normalized element $T_A \in \ker(A)$ by

(1.1) $AT_A = 0$, $||T_A|| = 1$,   and $\det\begin{pmatrix} A \\ T_A^* \end{pmatrix} > 0$.

Here $||.||$ denotes Euclidean norm and $(.)^*$ denotes transposition. The Moore-Penrose inverse A^+ of A is an $(N+1,N)$-matrix defined by

(1.2) $\begin{pmatrix} A \\ T_A^* \end{pmatrix}^{-1} = (A^+, T_A)$,

see e.g. Ben-Israel and Greville [2].

There are two flows which play an important role in curve tracing, c.f. Georg[6], namely

(1.3) $\dot{x} = T_{DH(x)}$

and

(1.4) $\dot{x} = -DH(x)^+ H(x)$.

Here $DH(x)$ denotes the Jacobian of H at x. The above flows are only defined for regular points $x \in \mathbb{R}^{N+1}$ of H and they are orthogonal to each other. The first goes back to Davidenko[4] and is often used to trace a curve $c(s)$ by numerical integration. The second is a continuous version of Newton's method for underdetermined systems of nonlinear equations, and has recently been studied by Tanabe[9]. For convergence properties of Newton's method related to (1.4), we refer to Ben-Israel[1]. It is easily seen that the property of local quadratic convergence holds also for underdetermined systems of equations such as those considered here.

(2) A PREDICTOR-CORRECTOR METHOD

The idea of Haselgrove[7] to numerically follow the curve $c(s)$ is to make a predictor step $x_{n+1,1}$ in accordance with points $x_n, \ldots, x_{n-k}$ previously found on the curve, and then start Newton's method at $x_{n+1,1}$ in order to obtain a new point x_{n+1} on the curve $c(s)$. To be specific, let us here suggest an Adams-Bashforth predictor, applied to the differential equation (1.3). This predictor is not so sensitive to the accuracy with which the previous points were obtained, since it only involves the tangents $T_{DH(x)}$ at the points $x=x_n, \ldots, x_{n-k}$ and hence has better stability properties than Hermite interpolation.

Let h_i be the stepsize by which the predicted point $x_{i,1}$ was obtained, and let

$$(2.1) \quad s_j := \sum_{i \leq j} h_i .$$

Consider the Adams-Bashforth predictor

$$(2.2) \quad x_{n+1,1} = x_n + \int_{s_n}^{s_{n+1}} P_{n,k}(s) \, ds$$

where $P_{n,k}(s)$ is the polynomial in s of degree k with coefficients in $\mathbb{R}^{N+1}$ which interpolates the tangent $T_{DH(x)}$ at the points $x=x_n, \ldots, x_{n-k}$. Subsequent Newton steps are then applied to "correct" $x_{n+1,1}$:

$$(2.3) \quad \begin{cases} x_{n+1,i+1} = x_{n+1,i} - DH(x_{n+1,i})^+ H(x_{n+1,i}) \\ x_{n+1} = \lim_{i \to \infty} x_{n+1,i} \end{cases}$$

A considerable amount of computational effort may be saved by using a rank-two update for $DH(x_{n+1,i})$, $i>1$. This update has been proposed by Georg[6] and is based on Broyden's method[3]. It has the advantage (crucial for the next predictor step) that the tangent $T_{DH(x_{n+1})}$ may be very precisely approximated.

(3) SOME SIMPLE ASYMPTOTIC ESTIMATES

Assuming that the predictor-corrector method (2) is performed with constant step size $h:=h_i$, some simple asymptotic estimates may be derived. Let us denote by $x(s_{n+1})$ the solution of (1.3) at s_{n+1}, beginning with the initial value $x(s_n)=x_n$. The local truncation error, see e.g. Shampine and Gordon[8], is given by

(3.1) $x_{n+1,1}-x(s_{n+1}) = O(h^p)$, $p=k+2$.

By Taylor's formula, since $H(x(s_{n+1}))=0$, we obtain

(3.2) $H(x_{n+1,1}) = O(h^p)$,

and consequently

(3.3) $x_{n+1,2}-x_{n+1,1} = O(h^p)$.

From $x_{n+1,2}-x_{n+1,1} = -DH(x_{n+1,1})^+ H(x_{n+1,1})$ we also conclude that the first two summands in

$$H(x_{n+1,2})=H(x_{n+1,1})+DH(x_{n+1,1})(x_{n+1,2}-x_{n+1,1})+O(h^{2p})$$

cancel, hence

(3.4) $H(x_{n+1,2}) = O(h^{2p})$.

Assuming that the estimates obtained above have non-zero leading terms, we find an estimate for the "contraction"

(3.5) $\kappa_{n+1} := \dfrac{||H(x_{n+1,2})||}{||H(x_{n+1,1})||} = O(h^p)$.

In the same way, we derive estimates for a "scaled contraction"

$$(3.6) \quad \bar{\kappa}_{n+1} := \frac{||DH(x_{n+1,1})^+ H(x_{n+1,2})||}{||DH(x_{n+1,1})^+ H(x_{n+1,1})||} = O(h^p) ,$$

and for the following two possibilities of measuring the "distance to the curve"

$$(3.7) \quad d_{n+1} := ||H(x_{n+1,1})|| = O(h^p) ,$$

$$(3.8) \quad \bar{d}_{n+1} := ||DH(x_{n+1,1})^+ H(x_{n+1,1})|| = O(h^p) .$$

A simpler argument shows that for successive angles the estimate

$$(3.8) \quad \alpha_{n+1} := arcos(T^*_{DH(x_n)} \ T_{DH(x_{n+1})}) = O(h)$$

holds.

(4)   A STEP SIZE CONTROL

The estimates (3) lead to a natural way of controling the step size h. Let us illustrate this by the contraction (3.5). Assuming an estimate of the form

$$(4.1) \quad \kappa_{n+1} = C_{n+1} h^p + O(h^{p+1})$$

we may "forget" the higher order term $O(h^{p+1})$. For a fixed stepsize h, we observe a contraction κ_{n+1} and thus measure approximately

$$(4.2) \quad C_{n+1} \cong \kappa_{n+1}/h^p .$$

If we want to maintain an "ideal contraction" κ_{ideal} (defined by the user) then the above formula suggests an ideal step size h_{ideal} by $\kappa_{ideal} = C_{n+1} h^p_{ideal}$ or

$$(4.3) \quad \begin{cases} h_{ideal} = \beta_{n+1} h \\ \beta_{n+1} = (\kappa_{ideal}/\kappa_{n+1})^{1/p} . \end{cases}$$

There are two ways of using (4.3) in order to get a
step size control:

(i)   A well known approach in numerical integration
is to try to keep the current step size h fixed and
adapt it only when the formula (4.3) suggests a
considerable change, say $\beta_{n+1} < 1/2$ or $\beta_{n+1} > 2$.
This not only saves computational effort for the
Adams-Bashforth formula, see e.g. Shampine and Gordon[8],
but also makes the motivation for the control (4.3) less
heuristic.

(ii)   Vary the step size in every step. The problem
is then how to interpret h and h_{ideal} in (4.3). A
careful recalculation of the estimates (3) for variable
stepsize leads to the following formula

$$(4.4) \quad \int_{s_{n+1}}^{s_{n+2}} (s-s_{n+1}) \ldots (s-s_{n+1-k}) \, ds =$$

$$\beta_{n+1}^{p} \int_{s_n}^{s_{n+1}} (s-s_n) \ldots (s-s_{n-k}) \, ds ,$$

which can be used to obtain the next step size
$h_{n+2} = s_{n+2} - s_{n+1}$. If one does not want to undertake this
somewhat complicated task, we propose an easier control
which comes close to (4.4), but is less well justified:

$$(4.5) \quad h_{n+2} = \beta_{n+1} h_{n+1} .$$

Note that by formulae such as (4.3), the control (4.5)
aims at every step to obtain the desired contraction
κ_{ideal} . If p increases, the control is damped which
is still tolerable. It would be much worse to under-
estimate the sensitivity of the control by using too
small p, since this would result in oscillations around
the right stepsize.

A more sophisticated stepsize control is due to
Deuflhard[5]. He strives to take the largest possible
step such that the predicted point still lies within the
region of attraction of Newton's method. This is, roughly
speaking, done by using the previously obtained data to
estimate the constants of the Newton-Kantorovitch theorem.
His method is very efficient for embedding methods
related to multiple shooting techniques.

The control strategy (4.3),(4.5) has the advantage
that it allows additional observable quantities besides
contraction (e.g. distance, angle, etc.) to also
determine the step size. The user decides how safe the
method should be by entering the "ideal" quantities for
those observables. In this way, we measure several
factors β_{n+1} corresponding to various observables, and
the control, e.g. (4.5), is then performed by taking the
minimum of these factors. Consequently, at different
parts of the solution curve, different observables
become active in the control. The result is a very safe
curve tracing algorithm which we found especially
useful for "nasty" solution curves.

(5)   EXAMPLE OF AN ALGORITHM

To illustrate what has been discussed, we finally
sketch a type of algorithm which we found quite efficient
and safe. It may be doubted whether a high order Adams-
Bashforth predictor is necessary, since Newton's method
is very powerful. Therefor, we use a low order (p=3)
Adams-Bashforth predictor, and the correcting Newton
steps are accompanied by a rank two update as mentioned
earlier.

Since the following rough description of an algorithm
serves just as an illustration, we want to avoid complicated technical details. Hence, we do not discuss how
the linear equations involved are solved in an efficient
and stable way, nor do we give any discussion on how
to organize successive operations in order to minimize
the numerical effort.

1   Remark: Start

2   Enter the following data:

x= point on the curve

c_i= ideal contraction

d_i= ideal distance

α_i= ideal angle

h_0= smallest allowed step size

  = desired accuracy

h_1= initial step size

3   Remark: Initialization

4   A:= Jacobian of H at x

5    $t_1 := t_2 := T_A$, c.f. (1.1)

6    $h_2 := h_1$

7   Remark: Initialize control factor

8    $\beta := 2$

9   Remark: Failure?

10   if $h_2 < h_0$ then print "failure" and stop

11   Remark: Predictor

12   $y := x + h_2 t_2 + \dfrac{h_2^2}{2h_1}(t_2 - t_1)$

13   A:= Jacobian of H at y

14   $t_3 := T_A$, c.f. (1.1)

15   $\beta := \min(\beta, (d_i/||A^+H(y)||)^{1/3}, \alpha_i/\arccos(t_2^* t_3))$

16   if $\beta < 1/2$ then goto 29

17 Remark: Newton step

18 $z := y - A^+ H(y)$

19 if $||y-z|| > h_o$ then

$$\beta := \min(\beta, (c_i/(||A^+H(z)||/||A^+H(y)||))^{1/3})$$

20 if $\beta < 1/2$ then goto 29

21 Remark: Newton update

22 if $||y-z|| > h_o$ then $A := A + \dfrac{H(z)(z-y)^*}{||z-y||^2}$

23 Remark: Tangential update

24 $u := z + h_0 t_3$

25 $A := A + \dfrac{(H(u)-H(z))t_3^*}{h_0}$

26 $t_3 := T_A$, c.f. (1.1)

27 Remark: Corrector finished?

28 if $||z-y|| > h_0$

then $y := z$ and goto 17

else $t_1 := t_2$, $t_2 := t_3$, $h_1 := h_2$, $h_2 := \beta h_2$, $x := z$
and goto 7

29 Remark: Repeat predictor with smaller step size

30 $h_2 := h_2/2$ and goto 7

REFERENCES

1. Ben-Israel,A.: A modified Newton-Raphson method for the solution of systems of equations. Israel J. Math.3 (1965) 94-98

2. Ben-Israel,A. and Greville,T.N.E.: Generalized inverses: theory and applications. Wiley-Interscience Publ., 1974

3. Broyden,C.G.: A class of methods for solving nonlinear simultaneous equations. Math.Comp.19 (1965) 577-593

4. Davidenko,D.: On a new method of numerical solution of systems of nonlinear equations. Doklady Akad. Nauk SSSR (N.S.) 88 (1953) 601-602

5. Deuflhard,P.: A stepsize control for continuation methods and its special application to multiple shooting techniques. Numer.Math.33 (1979) 115-146

6. Georg,K.: Numerical integration of the Davidenko equation. In:"Numerical solution of nonlinear equations",E.Allgower, K.Glashoff, H.-O.Peitgen (eds) Springer Verlag, Lecture Notes in Math.878 (1981)

7. Haselgrove,C.B.: The solution of non-linear equations and of differential equations with two-point boundary coditions. Comput.J.4 (1961) 255-259

8. Shampine,L.F. and Gordon,M.K.: Computer solution of ordinary differential equations: the initial value problem. W.H.Freeman and Comp., 1975

9. Tanabe,K.: Continuous Newton-Raphson method for solving an underdetermined system of nonlinear equations. Nonlinear Analysis, Theory, Methods and Applications 3 (1979) 495-503

ON A CLASS OF LINEAR COMPLEMENTARITY PROBLEMS OF VARIABLE DEGREE

Roger Howe

Mathematics Department
Yale University
New Haven, Connecticut

INTRODUCTION:

We consider the linear complementarity problem (LCP): Given an n-vector z and an $n \times n$ matrix M, find n-vectors x and y such that

a) $x - My = z$ b) $x \geq 0, \quad y \geq 0$

c) For each i, either $x_i = 0$ or $y_i = 0$.

In [ES] it was shown that the LCP amounted to the problem of inverting a piecewise linear map

$$(1.1) \qquad P_M : \mathbb{R}^n \to \mathbb{R}^n$$

The map P_M is linear on each orthant. Recall that an orthant is specified by requiring of an n-vector that each of its coordinates have a specified sign (if non-zero). Thus there are 2^n orthants. Let Q be an orthant. Define a matrix M_Q by the rule

(1.2)   a) If the j-th coordinate of vectors in Q is positive, the j-th column of M_Q is e_j, the j-th standard basis vector of $\mathbb{R}^n$.

      b)   If the j-th coordinate of vectors in Q is negative,
the j-th column of M_Q is the j-th column of M.

Then

(1.3) $$P_M | Q = M_Q$$

In [HS] it is shown how this interpretation of the LCP
allows one, under mild non-degeneracy assumptions to use
topological methods to obtain insight into the structure of the
LCP. In particular, under the condition

(WND) $$P_M(x) = 0 \Rightarrow x = 0$$

there is defined an integer $\deg(P_M)$, the <u>degree</u> of P_M, which
tells how many solutions, properly counted, there are to the LCP.
We will review this briefly.

      Observe that $\det M_Q$ is equal to the determinant of the
principal minor of M defined by taking the rows and columns of
M corresponding to the coordinates which are negative on Q. We
call this the Q-minor of M. Evidently M and M_Q have the
same Q-minor. Define an index by

(1.4) $$\mathrm{ind}(M,Q) = \begin{cases} 1 & \text{if} \quad \det M_Q > 0 \\ -1 & \text{if} \quad \det M_Q < 0 \end{cases}$$

If v is an n-vector with all coordinates non-zero, let $Q(v)$
be the orthant containing v. Then for a point z in general
position in $\mathbf{R}^n$, one has the formula

(1.5) $$\deg P_M = \sum_{P_M(v) \,=\, z} \mathrm{ind}\,(M,Q(v))$$

We remark also that the condition

(ND) $\det M_Q \neq 0$ for all Q

which is equivalent to requiring all principal minors of M to be
non-zero, implies (WND). Further, under condition ND, we can
define a local degree for P_M at all points of $\mathbb{R}^n$ (not merely
on the interior of orthants) such that the index formula (1.5)
holds not only for z in general position but for all z in $\mathbb{R}^n$.
(This extended formula is implicit in the analysis of §4 in [HS].)
Under the assumption ND, the collection { ind (M,Q)} of ± 1's
is called the sign pattern of P_M.

     The main point of [HS] was to illustrate how degree theory,
especially the index formula (1.5), allowed one to develop
systematically a variety of known results on LCP, and to extend
and refine these results. It was also pointed out that the main
classes of matrices M considered in connection with the
algorithmics of the LCP (see [C] for a discussion) yielded
maps P_M of degree 1; whereas there existed matrices M which
yielded maps P_M of very large degree, growing exponentially with
the dimension n of the problem. However, although some
examples and estimates of degree were given, no serious attack on
the difficult problem of actually computing the degree for given
matrices was attempted. The main purpose of this paper is to
study a class of matrices M for which deg P_M can assume many
values and to explicitly compute deg P_M for this class.
Corollary to this study, we obtain some results related to the
problem of Q-matrices, i.e. of describing those matrices M
such that P_M is surjective.

     It is a pleasure to acknowledge the stimulation I received
regarding this paper at the NATO Advanced Research Institute, held
in Porto Cervo, Sardinia on June 3-6, 1981, and organized by
Professor Floyd Gould. Conversations with L. Watson were
especially helpful.

2: Positive off-diagonal matrices

We will study the LCP for the family

(pod) $M = \{m_{ij}, m_{ij} \geq 0 \text{ if } i \neq j\}$

of matrices with positive off-diagonal entries. We note that the early example of Murty [M] of a matrix for which the LCP always has an even number of solutions, and its generalizations in [HS] belong to this class. Thus the present study is in part an attempt to understand these examples in a systematic way.

The positivity of the off-diagonal entries of M seems to affect the associated LCP in two different ways, which are expressed in corollary 2.2 and lemma 2.6.

Let $\mathbf{R}^{n+}$ denote the positive orthant, where all coordinates are non-negative.

Lemma 2.1: Let T be a matrix with non-positive off diagonal entries. Then either

i) $T(\mathbf{R}^{n+}) \supseteq \mathbf{R}^{n+}$,   or

ii) $T(\mathbf{R}^{n+})$ can be separated from (the interior of) $\mathbf{R}^{n+}$

Proof: Suppose first that some diagonal entry t_{ii} of T is non-positive. Then the i-th component of $T(x)$ will be non-positive for any x in $\mathbf{R}^{n+}$. Hence in this case the i-th coordinate function separates $T(\mathbf{R}^{n+})$ from $\mathbf{R}^{n+}$. Hence for possibility ii) to fail, all diagonal entries of T must be positive. Since diagonal matrices with positive diagonal entries map $\mathbf{R}^{n+}$ onto itself, we may multiply T by such a matrix so that the diagonal entries of T are all 1. Hence

$$T = I + T_0$$

where I is the identity matrix and T_0 has diagonal entries zero and non-positive off-diagonal entries. Thus $-T_0$ is a non-negative matrix. It is well known $[T]$ that there is a non-negative vector $v \neq 0$ such that

$$- v^t T_0 = \lambda v^t$$

where v^t denotes the transpose of v. Furthermore, the eigen-value λ is positive, and if $\lambda < 1$, then $T = I + T_0$ is invertible, with inverse represented by the series

$$T^{-1} = I - T_0 + T_0^2 - T_0^3 \ldots$$

which is clearly a non-negative matrix. Hence if $\lambda < 1$ we have

$$T^{-1}(R^{n+}) \subseteq R^{n+}.$$

Thus alternative i) holds. On the other hand, if $\lambda \geq 1$, we have

$$v^t T(x) = (1-\lambda) v^t x \leq 0 \quad -$$

for any $x \geq 0$. Hence in this case alternative ii) holds.
Since all possibilities have been covered, the lemma is proved.

Corollary 2.2: If the matrix M of the LCP is in (pod),
the class of positive off-diagonal matrices, then for any orthant
Q, either
     i) $P_M(Q) \supseteq R^{n+}$,   or
     ii) $P_M(Q)$ can be separated from the interior of) R^{n+}

Proof: Let R_Q denote the diagonal matrix with diagonal
entries $r_{ii} = \pm 1$, and such that $r_{ii} = +1$ exactly for those
coordinates which are positive on Q. It is easy to see that

$$R_Q^2 = I, \quad R_Q(R^{n+}) = Q, \quad R_Q(Q) = R^{n+}$$

Therefore

$$P_M(Q) = P_M R_Q(\mathbb{R}^{n+}) = M_Q R_Q(\mathbb{R}^{n+})$$

By inspection the matrix $M_Q R_Q$ has non-positive off-diagonal entries. Hence lemma 2.1 applies to yield the corollary.

It follows from corollary 2.2 that the problem of of computing $\deg P_M$ is considerably simpler than it might be. We do not have to look in detail at solutions of the LCP. We need only determine which orthants satisfy alternative i) of the corollary, and form sum of the $i(M,Q)$ for these Q. It turns out one can say a good deal about the structure of the set of these Q. The simplifying observation is the following

Lemma 2.3: Let T have non-positive off diagonal entries. Then any principal minor T' of T does also. If alternative ii) holds for T', it also holds for T.

Proof: Without loss of generality we may assume that T' is the leading $m \times m$ minor of T. Thus we can partition T

$$T = \begin{bmatrix} T' & T_{12} \\ T_{21} & T_{22} \end{bmatrix}$$

Let v be an m-vector such that

$$v^t T'(y) \leq 0 \qquad y \in \mathbb{R}^{m+}$$

Turn v into an n-vector by adding 0's for the last $n - m$ coördinates. Write a vector x in $\mathbb{R}^{n+}$ as $x = x_1 + x_2$ where x_1 has the last $n - m$ coordinates zero and x_2 has the first m coordinates zero. Then with T partitioned as above we have

$$v^t Tx = v^t T' x_1 + v^t T_{12} x_2 \leq 0$$

since $v^t T' x_1 \leq 0$ by choice of v and T_{12} is non-positive. Thus we see v also separates $T(\mathbb{R}^{n+})$ from $\mathbb{R}^{n+}$, and the lemma is proved.

To an orthant Q, let us attach the subset

(2.1) $E(Q) = \{ j : 1 \leq j \leq n : x_j \leq 0 \text{ for all } x \in Q \}$

Thus $E(\mathbb{R}^{n+}) = \emptyset$, the empty set, and $E(-\mathbb{R}^{n+})$ is the full set of integers from 1 to n. Set

(2.2) $\ell(Q) = {}^{\#}(E(Q))$, the cardinality of $E(Q)$.

For a matrix M in (pod), let $D(M)$ denote the set of orthants Q for which alternative i) of Corollary 2.2 holds.

Corollary 2.4: If M is in (pod) and $Q \in D(M)$, then also $Q' \in D(M)$ whenever $E(Q') \subseteq E(Q)$.

Proof: From the definitions we see that the Q'-minor of M is a minor of the Q-minor of M exactly when $E(Q') \subseteq E(Q)$. Since as we have noted the Q-minor of M is also the Q-minor of M_Q, the corollary follows directly from the lemma 2.3.

Theorem 2.5: Given a matrix M in (pod), and an orthant $Q \in D(M)$, we have the formula

(2.3) $\text{ind } (M,Q) = (-1)^{\ell(Q)}$

Conversely, we have $Q \in D(M)$ if (and only if) every Q' with $E(Q') \subseteq E(Q)$ satisfies (2.3). Thus if we define integers

(2.4) $\beta_j(M) = {}^{\#}(\{Q : Q \in D(M) \text{ and } \ell(Q) = j\})$

then

(2.5) $\deg P_M = \sum_{j=0}^{n} (-1)^j \beta_j(M)$

Proof: First observe that if alternative i) of corollary 2.2 holds for Q, then certainly M_Q is non-singular, hence

ind (M,Q) is defined. Formula (2.5) is immediate from (2.3) and (2.4). We prove formula (2.3) by induction on $\ell(Q)$. If $M \in \mathcal{D}(Q)$, by corollary (2.4) we can choose $Q' \in \mathcal{D}(Q)$ with $E(Q') \subseteq E(Q)$ and $\ell(Q') = \ell(Q) -1$. Then precisely one more coordinate is negative in Q than in Q'. Hence Q and Q' have a hyperplane face in common. Since both Q and Q' satisfy alternative i) of corollary (2.2), the cones $P_M(Q)$ and $P_M(Q')$ have points in common. Hence both $P_M(Q)$ and $P_M(Q')$ must lie on the same side of the hyperplane spanned by $P_M(Q \cap Q')$. It follows that P_M changes orientation as one crosses $Q \cap Q'$, or in other words

$$\text{ind } (M,Q) = - \text{ ind } (M,Q')$$

Since formula (2.3) holds for Q' by induction, it holds for Q.

To prove the condition on Q to belong to $\mathcal{D}(M)$, it suffices to consider the case when $Q = - R^{n+}$. For let M' be the Q-minor of M. It is easy to see from the form of M_Q that $M_Q(Q) \subseteq R^{n+}$ if and only if $M'(-R^{\ell+}) \subseteq R^{\ell+}$ where we have abbreviated $\ell(Q) = \ell$. So assume $Q = - R^{n+}$. Then our assumption simply amounts to $\text{ind}(M,Q') = (-1)^{\ell(Q')}$ for all orthants Q'. Thus M has the same sign pattern as $-I$. It follows from [HS], Theorem 4.5 that $\deg P_M = 0$. Since we may assume by induction on $\ell(Q)$ that our condition is true for all Q' with $\ell(Q') < n = \ell(- R^{n+})$, the images of all these Q' by P_M cover R^{n+}. For the index formula (1.5) to hold for $z \in R^{n+}$, it is necessary that $P_M(-R^{n+})$ also. This finishes the proof of Theorem 2.5.

Remark: Thus within the case (pod) the sign pattern of M already determines $\deg P_M$. First an orthant Q is the $\mathcal{D}(M)$ if and only if all orthants Q' with $E(Q') \subseteq E(Q)$ satisfy equation (2.3). Then $\deg (P_M)$ is determined by formula (2.5). By contrast, already for 2×2 matrices, the sign pattern does not determine the degree of a general matrix. Theorem 2.5 can be used fairly effectively to compute $\deg P_M$ for specific M in (pod). See section 3 for some examples.

The other feature of positive off-diagonal entries is that they permit an improvement of lemma 4.8 of [HS]. To state this we recall the concepts involved in that lemma. A semi-orthant S is the subset of R^n defined by specifying the signs of some of the coordinates. Let $N = \{1,2,\ldots,n\}$ denote the set of integers from 1 to n. Let $J \subseteq N$ be a subset of N and let N - J denote the complementary subset so that N is the disjoint union of J and N - J. Let R_J^+ denote the semi-orthant on which the coordinates x_j, with $j \in J$, are non-negative. We call R_J^+ the positive semiorthant defined by J. Clearly an orthant Q is contained in R_J^+ if and only if $E(Q) \subseteq N-J$. Define subspaces

$$(2.6) \qquad V_1(J) = \{x \in R^n : x_\ell = 0 \text{ if } \ell \notin J\}$$

$$V_2(J) = \{x \in R^n : x_j = 0 \text{ if } j \in J\}$$

Then we have an orthogonal direct sum decomposition

$$\mathbb{R}^n = V_1(J) \oplus V_2(J)$$

The space $V_2(J)$ is contained in $\mathbb{R}_J^+$, and is called the spine

of $\mathbb{R}_J^+$. The set $Q_J^1 = \mathbb{R}_J^+ \cap V_1(J)$ is clearly an orthant in $V_1(J)$.

We have the decomposition

$$\mathbb{R}_J^+ = Q_J^1 + V_2(J)$$

Let M be an $n \times n$ matrix, not necessarily in (pod). If

J, J' are subsets of N, let $M_{J,J'}$ denote the minor of M whose

rows are chosen from J and whose columns are chosen from J'.

If $J' = J$, then $M_{J,J}$ is a principal minor of M. If $J = N$

and $J' = \{\ell\}$, a singleton, then $M_{J,J'}$ is the ℓ-th column of M.

Assume that M satisfies condition (ND). Choose a subset

$J \subseteq N$. As explained in [HS], the minor $M_{N-J,N-J}$ of M defines

in an natural way a LCP on $V_2(J)$ whose matrix is $M_{N-J,N-J}$.

The associated map $P_{M_{N-J,N-J}} = P_M(J)$ is called the local map

around Q_J^1 defined by P_M.

Lemma 2.6: Suppose the minor $M_{J,N-J}$ of M has positive

entries. Suppose also that $M_{N-J,N-J}$ is a Q-matrix, i.e., the

local map $P_M(J)$ is surjective from $V_2(J)$ to $V_2(J)$. Then

(2.7) $P_M(\mathbb{R}_J^+) \supseteq \mathbb{R}_J^+$

Proof: Suppose $P_M(J)$ is surjective. This means that,

given $x \in V_2(J)$, we can find a subset $K \subseteq N - J$ and non

negative numbers $\{a_k : k \in K\}$ and $b_\ell : \ell \in N-J-K$ such that

(2.8) $$x = \Sigma a_k e_k - \Sigma b_\ell M_{N-J, \{\ell\}}$$

Here again, the e_k are the standard basis vectors of R^n, the
columns of the identity matrix. It is understood that the column
$M_{N-J, \{\ell\}}$ defines a vector in $V_2(J)$ by letting all its coordinates
in J be zero.

Consider $z \in R_J^+$. Write $z = y + x$ with $y \in Q_J^1$ and
$x \in V_2(J)$. Consider the difference

$$z - (\Sigma a_k e_k - \Sigma b_\ell M_{N, \{\ell\}}) = y + \Sigma b_\ell M_{J, \{\ell\}} = y_1$$

Equation (2.8) tells us that $y_1 \in V_1(J)$. The assumption that
$M_{J,N-J}$ has positive entries implies that $y_1 \in Q_J^1$. In other
words for some non-negative numbers $\{a_j : j \in J\}$ we have

$$z = \Sigma a_j e_j + \Sigma a_k e_k - \Sigma b_\ell M_{N,\ell}$$

This equation implies $z \in P_M(R_J^+)$ as claimed; the lemma is proved.

Remark: This result should be of direct use in constructing
algorithms to solve the LCP for M in (pod). We note that the
fact that, if M is in (pod) and is not a positive matrix then
P_M folds several orthants over R^{n+}, will make it difficult to
construct Lemke-type algorithms for M. (See the remarks on
algorithms in [HS], section 6). We will see how lemma 2.6 can
be used on examples in section 3.

§3: Q-matrices and superfluous matrices.

The question of which matrices M are Q-matrices, i.e. are such that P_M is surjective, so that the LCP for M always has a solution, has been much studied (c.f. [C]). From the formula (1.5) one sees that the LCP always has at least $|\deg P_M|$ solutions. In particular if $\deg P_M \neq 0$, then M is a Q-matrix. This was observed in [HS]. The question naturally arises if the converse holds. In fact, before [HS] was written, Kelly and Watson [KW] had given an example which implies the converse fails. The interpretation of the Kelly-Watson example in the context of degree theory was given by Garcia and Gould [GG].

If M is a Q-matrix of degree zero, then for every point z in general position in R^n, there will be cancellation in the in the index formula (1.5), so there are more solutions to $P_M(v) = z$ than are necessary to account for $\deg P_M$. We can imagine the same thing happening for a general matrix. Thus let us define M to be <u>superfluous</u> if there are always more solutions to $P_M(v) = z$ than the minimum $|\deg P_M|$ i.e., if there is always cancellation in the index formula (1.5). Then if $\deg P_M = 0$ for some matrix M, then M is a Q-matrix if and only if M is superfluous. The notion of superfluity is related to the viability of Lemke-type algorithms. It was explained in [HS], section 6 that starting a path-following algorithm at a point z such that $P_M^{-1}(z)$ consists of one point guarantees its success. Of course if $P_M^{-1}(z)$ is a singleton, then $|\deg P_M| = 1$. If $|\deg P_M| > 1$, then refinements of the discussion in [HS] show that if ${}^\#(P_M^{-1}(z)) = |\deg P_M|$, then a path-following algorithm starting at z will succeed. Hence superfluous matrices are unfortunate because they allow no natural starting place for Lemke-type algorithms. It would be pleasant if they did not exist. Unfortunately they do. The Kelly-Watson example shows superfluous matrices M with

with deg P_M = 0 exist. We will show here that superfluous
matrices M with deg P_M arbitrary exist.

We begin with 2 examples from the class (pod). The first
is a superfluous matrix M_1 with deg P_{M_1} = 0. It is perhaps
easier to understand than the Kelly-Watson example. The second
matrix M_2 is 5 × 5 with deg P_{M_2} = 1, and it is also
superfluous.

Example 3.1. Consider the matrix

$$M_1 = \begin{bmatrix} -1 & a & a & b \\ a & -1 & a & b \\ a & a & -1 & b \\ b & b & b & -1 \end{bmatrix}$$

where a, b satisfy the inequalities

(3.1) $\frac{1}{2} < a < 1 < b < 2$

Then deg P_{M_1} = 0, but M_1 is a Q-matrix.

Proof: To compute deg P_{M_1} we use Theorem 2.5. We will
indicate the orthants Q of $\mathbb{R}^4$ by their E(Q) of formula (2.1).
Thus Q($\emptyset$) = $\mathbb{R}^{4+}$, and Q(1,3) is the orthant such that
x_1, x_3 $\le$ 0 $\le$ x_2, x_4. Computing the determinants of the principal
minors of M_1 and taking inequalities (3.1) into account we get the
following table. Here we have abbreviated deg(M_1, Q(1,3)) = d(1,3)
and so forth.

(3.2) $d(\emptyset) = 1$

$$d(1) = -1, d(2) = -1, d(3) = -1, d(4) = -1$$

$$d(1,2) = 1, d(1,3) = 1, d(1,4) = -1, d(2,3) = 1, d(2,4) = -1, d(3,4) = -1$$

$$d(1,2,3) = 1, d(1,2,4) = 1, d(1,3,4) = 1, d(2,3,4) = 1$$

$$d(1,2,3,4) = -1$$

From table (3.2) and Theorem 2.5 we read off

(3.3) $\mathcal{D}(M) = \{Q(\emptyset), Q(1), Q(2), Q(3), Q(4), Q(1,2), Q(1,3), Q(2,3)\}$

Hence from formula (2.5), we get

$$\deg P_{M_1} = 1 - 4 + 3 = 0.$$

On the other hand, if we compute the degrees of the maps associated to the 3×3 principal minors of M_1 we find they are either 1 or -1. (All these computations are facilitated by the fact that M_1 is unchanged by permutations of the first 3 coordinates in $\mathbb{R}^4$. For example, we need only to compute for 2 of the 4 3×3 principal minors.) Thus the 3×3 minors of M_1 are certainly Q-matrices. It follows from lemma 2.6 that

$$P_{M_1}(\mathbb{R}^+_{\{i\}}) \supseteq \mathbb{R}^+_{\{i\}}$$

for each half-space $\mathbb{R}^+_{\{i\}}$, $i = 1, 2, 3, 4$: Hence to finish verifying that M_1 is a Q-matrix, it suffices to show that the image of P_{M_1} contains the negative orthant $-\mathbb{R}^{4+}$.

Consider the image by P_{M_1} of the orthant $Q(1,2,3)$. We have

$$M_{1Q}(1,2,3) = \begin{bmatrix} -1 & a & a & 0 \\ a & -1 & a & 0 \\ a & a & -1 & 0 \\ b & b & b & 1 \end{bmatrix}$$

With R_Q as corollary 2.2 we have

$$P_{M_1}(Q(1,2,3)) = M_{1Q(1,2,3)}(Q(1,2,3)) = M_{1Q(1,2,3)}R_{Q(1,2,3)}(\mathbb{R}^{4+})$$

We compute

$$(M_{1Q(1,2,3)}R_{Q(1,2,3)})^{-1} = \begin{bmatrix} c & d & d & 0 \\ d & c & d & 0 \\ d & d & c & 0 \\ e & e & e & 1 \end{bmatrix} = T$$

where

$$c = \frac{a-1}{(a+1)(2a-1)} \;,\; d = \frac{-a}{(a+1)(2a-1)} \;,\; e = \frac{-b}{2a-1}$$

Set

(3.4) $$z = -\sum_{i=1}^{4} z_i e_i$$

with $z_i \geq 0$. Then we compute that $T(z)_i > 0$ for $i = 1, 2, 3$, and

$$T(z)_4 = \frac{b(z_1+z_2+z_3)}{2a-1} - z_4$$

Hence in order that z belong to $P_{M_1}(Q(1,2,3))$ it is necessary and sufficient that

(3.5) $$z_4(2a-1) \leq b(z_1 + z_2 + z_3)$$

Next consider the image of $Q(3,4)$ under P_{M_1}. We have

$$M_1 Q(3,4) = \begin{bmatrix} 1 & 0 & a & b \\ 0 & 1 & a & b \\ 0 & 0 & -1 & b \\ 0 & 0 & b & -1 \end{bmatrix}$$

As with $Q(1,2,3)$ we have

$$P_{M_1}(Q(3,4)) = M_{1Q(3,4)}(Q(3,4)) = M_{1Q(3,4)} R_{Q(3,4)} (R^{4+})$$

We compute

$$(M_{1Q(3,4)} R_{Q(3,4)})^{-1} = \begin{bmatrix} 1 & 0 & h & \ell \\ 0 & 1 & h & \ell \\ 0 & 0 & f & g \\ 0 & 0 & g & f \end{bmatrix} = T'$$

where

$$(3.6) \quad f = \frac{-1}{b^2 - 1} \quad g = \frac{-b}{b^2 - 1} \quad h = \frac{-(a+b^2)}{b^2 - 1} \quad \ell = -\frac{b(a+1)}{(b^2 - 1)}$$

Taking z as in equation (3.4) and applying T' we find $T'(z)_3$ and $T'(z)_4$ are alway ≥ 0. We have

$$T'(z)_1 = -z_1 + \frac{(a+b^2)z_3 + b(a+1)z_4}{b^2 - 1}$$

$$T'(z)_2 = -z_2 + \frac{(a+b^2)z_3 + b(a+1)z_4}{b^2 - 1}$$

Hence for z to be in $P_{M_1}(Q(3,4)$ it is necessary and sufficient that

$$(3.7) \qquad \max (z_1, z_2) \le \frac{(a+b^2) z_3 + b(a+1) z_4}{(b^2-1)}$$

under our conditions (3.1) on a and b we have

$2a -1 < 1 < b$. Hence inequality (3.5) will hold if

$z_4 \le z_1 + z_2 + z_3$. On the other hand, we have $a + 1 > 3/2$, so

that $b(a+1) > b^2-1$ as long as $b < 2$. Hence if

$z_4 > z_1 + z_2 + z_3$, the inequality (3.7) obviously holds. Thus

we have shown that $-R^{4+}$ is in the union $P_{M_1} (Q(1,2,3) \cup Q(3,4))$

and example 3.1 is established.

Example 3.2: Consider the matrix

$$M_2 = \begin{bmatrix} -1 & a & a & b & b \\ a & -1 & a & b & b \\ a & a & -1 & b & b \\ b & b & b & -1 & b \\ b & b & b & b & -1 \end{bmatrix}$$

where a, b satisfy the inequalities (3.1).

Then $\deg P_{M_2} = -1$, but the equation $P_{M_2} (v) = z$ always has at

least 2 (hence 3) solutions, so that M_2 is superfluous.

Proof: Again we compute the sign pattern of M_2. We use

notation similar to that of Example 3.1. We obtain

$$(3.8) \qquad d(\emptyset) = 1$$

$$d(1) = -1, d(2) = -1, d(3) = -1, d(4) = -1, d(5) = -1$$

$$d(1,2)=1, d(1,3)=1, d(1,4)=-1), d(1,5)=-1, d(2,3)=1, d(2,4)=-1, d(2,5)=-1$$

$$d(3,4) = -1, \quad d(3,5) = -1, \quad d(4,5) = -1$$

$$d(i,j,k) = 1 \quad \text{for all} \quad \text{3-tuples} \quad i,j,k.$$

From table (3.8) we can read off that

(3.9) $(M_2) = \{Q(\emptyset), Q(1), Q(2), Q(3), Q(4), Q(5), Q(1,2), Q(1,3), Q(2,3)\}.$

Hence

$$\deg P_{M_2} = 1 - 5 + 3 = -1$$

To show that M_2 is superfluous, we must show for each point $z \in \mathbb{R}^n$ that either

(3.10)   i) $P_{M_2}^{-1}(z)$ contains at least 2 points, or

         ii) $P_{M_2}^{-1}(z)$ contains a point at which the local index
            is +1.

Consider the 4×4 minors of M_2. The minors obtained by eliminating either the 4th or 5th row and column of M_2 are precisely M_1. From the sign pattern (3.8) of M_2, we see that the minors obtained by eliminating the 1st, 2nd or 3rd row and column have index -2. It follows from the proof of lemma 2.6 and example 3.1 that the points in the halfspaces $\mathbb{R}_{\{4\}}^+$ and $\mathbb{R}_{\{5\}}^+$ satisfy either of criteria (3.10), while the points in the halfspaces $\mathbb{R}_{\{j\}}^+$ with $j = 1, 2,$ or 3 satisfy criterion (3.10) i). Thus we come again to consideration of the negative orthant.

Consider the image of $Q(1,2,3)$ under P_{M_2}. The computations here are very similar those for $Q(1,2,3)$ in example 3.1. We find that if

(3.11) $$z = - (\sum_{i=1}^{5} z_i e_i) \qquad z_i \geq 0$$

then   z   is $P_{M_2}(Q(1,2,3))$ if and only if

(3.12) $\max(z_4,z_5)(2a-1) \leq b(z_1+z_2+z_3)$

Thus all   z   for which   (3.12)   holds satisfy criterion   (3.10 ii).

Next consider the image of $Q(3,4)$.   We have

$$M_{2_{Q(3,4)}} = \begin{bmatrix} 1 & 0 & a & b & 0 \\ 0 & 1 & a & b & 0 \\ 0 & 0 & -1 & b & 0 \\ 0 & 0 & b & -1 & 0 \\ 0 & 0 & b & b & 1 \end{bmatrix}$$

We compute

$$M_{2_{Q(3,4)}} R_{Q(3,4)}^{-1} = \begin{bmatrix} 1 & 0 & h & \ell & 0 \\ 0 & 1 & h & \ell & 0 \\ 0 & 0 & f & g & 0 \\ 0 & 0 & g & f & 0 \\ 0 & 0 & m & m & 1 \end{bmatrix}$$

Here   f, g, h,   and ℓ are as equation   (3.6), and

$$m = \frac{-b}{b-1}$$

Thus, in order that   z   as in   (3.11) should be in $P_{M_2}(Q(3,4))$,

the inequalities   (3.7)   must hold.   Additionally, we must have

(3.13) $b(z_3+z_4) \geq (b-1)z_5$

Suppose $z_4 \geq z_5$. Then inequality (3.12) reduces to (3.5).
If it fails, then as in example 3.1 the inequalities (3.7)
follow. Further, the assumption $z_4 \geq z_5$ immediately implies
inequality (3.13). Therefore if $z_4 \geq z_5$, and z is not in
$P_{M_2}(Q(1,2,3))$, it is in $P_{M_2}(Q(3,4))$. Since the matrix M_2 is
invariant under permutations of the first 3 coordinates, such
z are also in $P_{M_2}(Q(1,4)$ and $P_{M_2}(Q(2,4))$. Thus these z
satisfy criterion (3.10) i). Since M_2 is also invariant under
permutation of the last 2 coordinates, the same conclusion holds
if $z_5 \geq z_4$. This covers all possibilities, so example 2 is
established.

From examples 3.1 and 3.2 by the direct sum technique of
[HS], section 5, we can construct very general superfluous
matrices. Before stating the result, let us refine slightly the
notion of superfluity. We will say a matrix M is <u>k-fold</u>
<u>superfluous</u> if for every $z \in \mathbb{R}^n$, the LCP associated to M has
at least $2k + |\deg P_M|$ solutions.

Theorem 3.3: Given integers d and $k > 0$, there exists in
sufficiently high dimensions a k-fold superfluous matrix M with
$\deg P_M = d$. Moreover these matrices may be taken to be of type
(pod).

Proof: Given an $n' \times n'$ matrix M' and an $n'' \times n''$ matrix
M'', we can form the direct sum matrix

$$M = \begin{bmatrix} M' & 0 \\ 0 & M'' \end{bmatrix}$$

which is an $(n'+n'') \times (n'+n'')$ matrix. As stated in [HS],
section 3, one has $\deg P_M = \deg P_{M'} \deg P_{M''}$. Furthermore, if $z = z'$
$z = z' + z''$ with $z' \in \mathbb{R}^{n'}$ and $z'' \in \mathbb{R}^{n''}$, it is easy to see that

$$P_M^{-1}(z) = P_{M'}^{-1}(z') \times P_{M''}^{-1}(z'')$$

Therefore if $M'(M'')$ is $k'(k'')$-fold superfluous, an easy calculation shows that M is k-fold superfluous, where

$$k = k' \deg P_{M''} + k'' \deg P_{M'} + 2k'k''$$

It is clear from Theorem 2.5 that we can find a matrix M_0 of type (pod) whose associated map has given degree. If this degree is non-zero, then the discussion above shows that if we take the direct sum of M_0 with an even number of copies of M_2 of example 3.2, we will produce a matrix of the same degree but of an arbitrarily many-fold superfluity. On the other hand, taking the direct sum of a large number of copies of M_1 of example 3.1 with itself, we produce a matrix of degree 0 and many-fold superfluity. This concludes Theorem 3.3.

Remark: In fact, just as in proposition 5.3 of [HS] we can make the foldness of superfluity grow exponentially with the dimension of the LCP.

Theorem 3.3 tends to reinforce the pessimism engendered by [HS] concerning the complexity of the LCP. We would like to finish with a result with somewhat the opposite tendency. Roughly speaking it says that if the sign pattern of a matrix is not too complicated, the matrix cannot be superfluous. Thus it is a qualitative result in the same spirit of the result of Murty et al [M] [STW] concerning matrices with only $+1$'s in their sign patterns.

Theorem 3.4. Let M be an $n \times n$ k-fold superfluous matrix, $k > 0$, and satisfying (ND). Suppose for simplicity that $\deg P_M \geq 0$. Then there must be at least $2(k+\deg P_M) + n - 1$ orthants Q with $\text{ind}(M,Q) = +1$, and at least $2k + n - 1$

orthants Q with ind $(M,Q) = -1$.

Remark: Suppose one wants to find a 3×3 Q-matrix of degree 0. Then by the theorem, one needs at least 4 orthants with positive index, and 4 with negative. But in R^3, there are only 8 orthants, so there must be exactly 4 of each. This can be made the basis of a proof that no 3×3 degree 0 Q-matrices exist. However, since this fact has already been established by Kelly-Watson [KW] and Garcia-Gould [GG], we will not pursue the argument.

Proof: If M is as specified, then R^n must be covered at least $k + \deg P_M$ times by the convex cones $P_M(Q)$ where Q is an orthant with ind $(M,Q) = +1$. Similarly, the cones $P_M(Q')$ with ind $(M,Q') = -1$ must cover R^n at least k times. Thus the theorem will follow from the next lemma.

Lemma 3.5: If $\{C_i\}_{i=1}^{\ell}$ is a collection of proper, closed pointed convex cones in R^n which cover R^n at least $k > 0$ times, in the sense that each point in R^n belongs to at least k cones, then $\ell \geq 2k + n - 1$.

Proof: The basic case is $k = 1$; this is basically a topological fact, but there is an easy geometric proof. Take cone C_1. Since it is closed & pointed, we can find a hyperplane $V \subseteq R^n$ such that

$$V \cap C_1 = \{0\}$$

Put $C_i' = V \cap C_i$ for $i \geq 2$. Then the C_i' must cover V. By induction on dimension, there must be at least n of the C_i'; adding C_1 gives at least $n + 1$ of the C_i. Moreover, this proof when carried on, implies that any $n - 1$ cones will leave uncovered at least one full line in R^n. A given pointed cone can cover at most one of the two rays emanating from the origin along this line. Hence to cover this line k times requires at

least 2k further cones, equally divided among the 2 rays. Thus
in all we need 2k + n - 1 as stated.

Corollary: An n × n Q matrix M of degree 0 must have a
sign pattern containing at least (n+1) = 1's and (n+1) -1's.

Proof: This is immediate from the theorem.

References

[C]   R. Cottle, Completely-Q Matrices, Stanford Univ. Dept. of
        Op. Res. Technical Report 79-12, Sept. 1979.

[ES]   B. C. Eaves and H. Scarf, The Solution of Systems of
        Piecewise Linear Equations, Math. Op. Res. 1, (1976), 1-27.

[GG]   C. Garcia and F. Gould, Studies in Linear Complementarity,
        in these Proceedings.

[HS]   R. Howe and R. Stone, Linear Complementarity and the Degree
        of Mappings, this volume.

[KW]   L. Kelly and L. Watson, Q-matrices and Spherical Geometry,
        Lin. Alg. and App. 25 (1979), 175-189.

[M]   K. G. Murty, On the Number of Solutions to Complementarity.
        Problems and the Spanning Properties of Complementary Cones,
        Lin. Alg. and App. 5 (1972), 65-108.

[STW]   H. Samuelson, R. M. Thrall, and O. Wesler, A Partition
        Theorem for Euclidean n-space, P.A.M.S. 9 (1958), 805-807.

[T]   A. Takayama, Mathematical Economics, Arden Press, Hinsdale,
        Ill., 1979.

LINEAR COMPLEMENTARITY AND THE DEGREE OF MAPPINGS*

Roger Howe

Richard Stone

Mathematics Department
Yale University
New Haven, Connecticut

Harvard University
Graduate School of Business
  Administration
Boston, Massachusetts 02163

INTRODUCTION

Let M be an $n \times n$ real matrix and q an n-vector. The problem: Find n-vectors x and y such that

(1a) $x - My = q$

(1b) $x \geq 0$, $y \geq 0$

(1c)   Either $x_i = 0$ or $y_i = 0$ for $1 \leq i \leq n$

is called the linear complementarity problem. As is explained in [C-D], several significant mathematical programming problems can be formulated as linear complementarity problems. For that reason, linear complementarity has been the subject of a considerable literature (see [K], [M], [L2], [G] and the papers cited there). For the most part in this literature the problem is treated from the algorithmic point of view, with the specification of procedures for solving the problem under various assumptions on the matrix M as a major goal. An exception to this rule is the paper [E-S] of Eaves and Scarf. In that paper, a general class of algorithms is discussed from a geometric point of view, and the linear

---

*This work was partially supported by NSF Grant MCS-79-05018.

179

complementarity problem is given a geometric interpretation, making
it amenable, for certain M, to the general methods of the paper.
The purpose of the present paper is to pursue the investigation of
linear complementarity from a geometric point of view[*]. Specifically, it will be shown that the topological theory of mapping
degree (as exposed in, say, [G-P]) has direct bearing on the
problem. No new algorithms will be proposed, but it is hoped the
considerations here will lend insight into how and why existing
algorithms work, and what can be expected of them.

In Section 2, the geometrization of the linear complementarity
problem is reviewed. This relates the problem to a mapping of the
(n-1)-sphere S^{n-1} to itself. In Section 3 we review the notion
of the degree of a mapping from S^{n-1} to itself, and recite some
of the basic properties of degree, in particular, its relation to
the more familiar notion of index. In Section 4 we apply the degree
theory to the maps coming from the linear complementarity problem.
This allows us to recapture quickly many of the known results on
the problem, including results of Eaves [E], of Murty [M], and of
Kojima-Saigal [K-S]. In particular, degree theory immediately
explains the widely noted fact that, under appropriate non-degeneracy assumptions, the parity of the number of solutions of (1) for
fixed M and a variable q is constant. Also, the degree is
relevant to the study of the class Q (see [C], [L2]) of matrices
M such that (1) always has a solution for any vector q, since if
the map associated to M has non-zero index, then automatically
M is in class Q . Section 5 studies the behavior of the degree
as a function of M. It is shown that the degree can grow exponentially in n, the dimension of the problem. By contrast, most

---

[*]By "geometric" we actually mean what might more precisely be called
"topological". Other papers which discuss geometric aspects of
linear complementarity are [C-R-S], [D-L], [G-G], [K-W], [Sa1],
[Sa2], [St], and [W]. Some of these were unknown to the authors at
the time of first writing.

algorithms work with matrices whose associated degree is 1. In Section 6 implications of this approach for algorithms are discussed, and some natural questions left unanswered here are given.

The first author would like to acknowledge the stimulation received from a talk by L. van der Heyden at the School of Organization and Management at Yale University, and conversations with H. Scarf. Both authors are indebted to R. W. Cottle for inspiration and encouragement.

2. GEOMETRIZATION OF THE LINEAR COMPLEMENTARITY PROBLEM

The basic construction of this section may be found in [E-S]. By an orthant in $\mathbb{R}^n$, we mean the closure of a convex cone where all coordinates have a fixed sign. Thus $\mathbb{R}^{n+}$, the positive orthant, is the cone in which all coordinates are non-negative. Clearly there are 2^n orthants all together. Given a matrix M, we will construct a piecewise linear map

$$(2) \qquad\qquad P_M: \mathbb{R}^n \to \mathbb{R}^n.$$

The map P_M will be linear on each orthant. Let e^i be the ith standard basis vector, and let m^i be the vector which defines the ith column of M. Let Q be an orthant. Define a matrix M_Q by the recipe:

(3a)   If x_i, the ith coordinate of $x \in Q$, is ≥ 0, then the ith column of M_Q is e^i,

(3b)   If $x_i < 0$ on Q, then the ith column of M_Q is m^i.

Then define P_M by:

(4)   If $x \in Q$, then $P_M(x) = M_Q x$.

It is easy to check that if x belongs to more than one orthant, that is, if some coordinates of x are zero, then $M_Q x$ does not depend on which Q we might choose, so that $P_M(x)$ is well-defined. This also implies that the mappings defined by the M_Q agree wherever orthants intersect so that P_M is continuous.

It is clearly linear on each orthant, hence piecewise linear overall. An alternative definition of P_M is as follows. For $x \in \mathbb{R}^n$, define $|x|$, the "absolute value" of x by

$$(5) \qquad |x| = (|x_1|, |x_2|, \ldots, |x_n|)$$

where $|x_i|$ denotes the usual absolute value of the ith coordinate of x. It is obvious that $|x|$ is continuous and linear on each orthant. (Also, $|x| = P_{(-I)}(x)$, where I is the identity matrix.) It is not hard to see that

$$(6) \qquad P_M(x) = \frac{1}{2}(x + |x|) - \frac{1}{2}M(|x| - x) = \frac{1}{2}(x + |x|) + \frac{1}{2}M(x - |x|)$$

Some pictures of P_M in the 2-dimensional case are in Figure 1.

We next observe that solving the linear complementarity problem (1) is equivalent to inverting P_M. For let x and y be vectors solving (1). Put $z = x-y$. Then $|z| = x+y$, and by comparison with (6) we find that (1) simply says $P_M(z) = q$. This establishes one direction of the first lemma. The other direction is just as easy.

<u>Lemma 2.1</u> (Eaves-Scarf). The linear complementarity problem (1) is equivalent to the problem:

Given the n-vector q, find z such that

$$(7) \qquad P_M(z) = q. \qquad\qquad \blacksquare$$

Since P_M is linear on orthants, it is in particular positive homogeneous of degree 1. That is,

$$P_M(\lambda x) = \lambda P_M(x)$$

for non-negative numbers λ. Let S^{n-1} be the unit sphere in $\mathbb{R}^n$. That is, S^{n-1} is the set of vectors x such that the Euclidean norm

$$\|x\| = (\sum_{i=1}^{n} x_i^2)^{1/2}$$

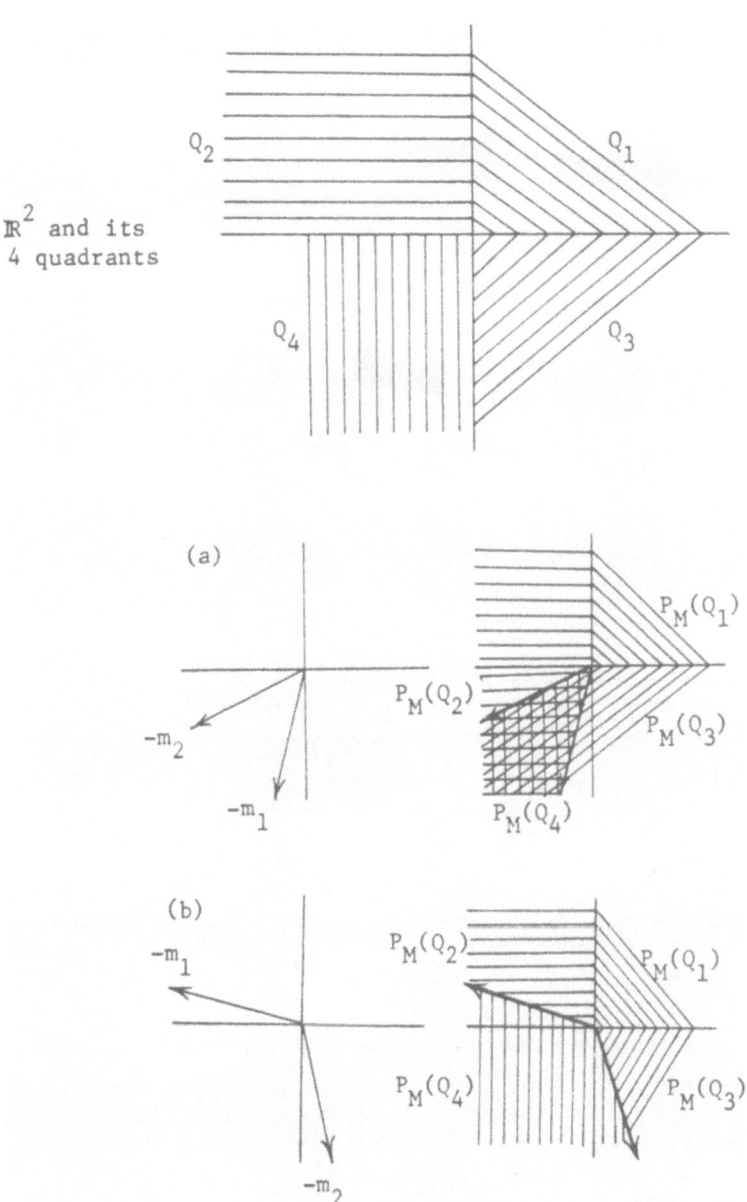

$\mathbb{R}^2$ and its
4 quadrants

FIGURE 1

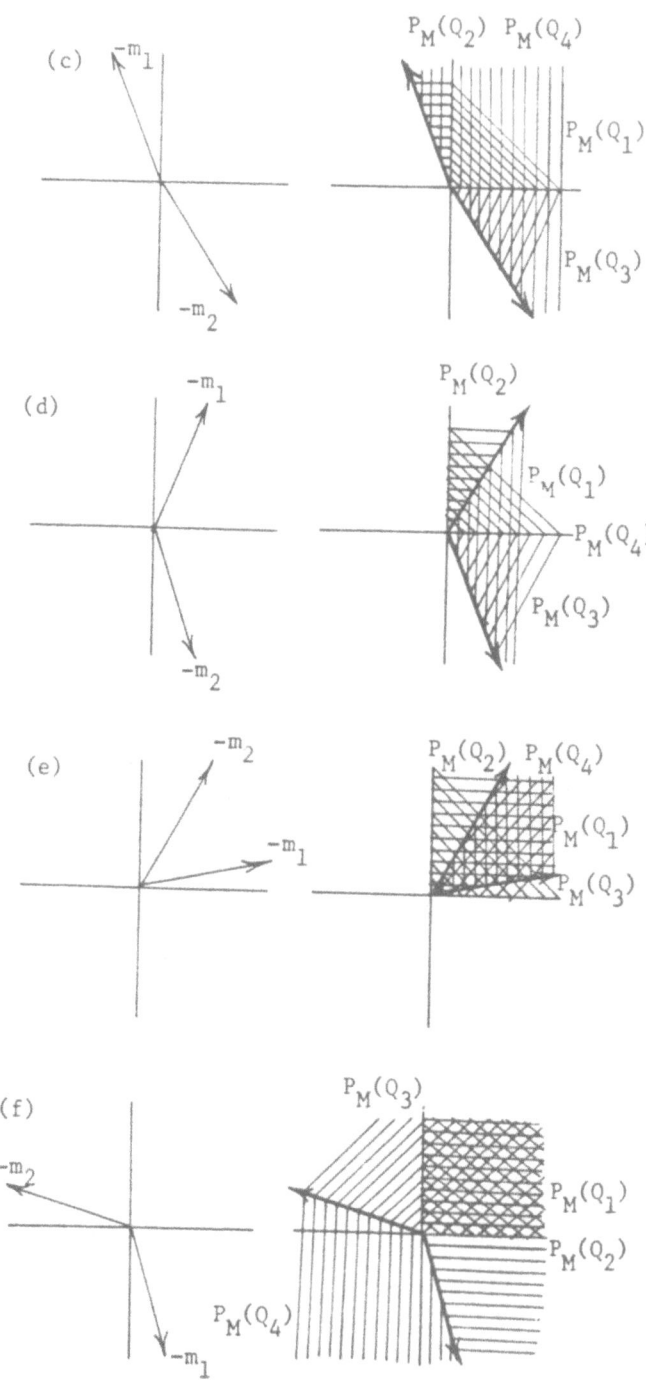

FIGURE 1 (continued)

equals 1. If $P_M(x) \neq 0$ whenever $x \neq 0$, we may define a map

$$\Pi_M : S^{n-1} \rightarrow S^{n-1}$$

by the formula

(8) $\qquad\qquad \Pi_M(x) = \|P_M(x)\|^{-1} P_M(x), \ x \in S^{n-1}$

We will refer to the intersection of S^{n-1} with an orthant in $\mathbb{R}^n$ as an orthant in S^{n-1}. Clearly Π_M is smooth (in fact, analytic) on each orthant of S^{n-1}.

When will it be true that $P_M(x) \neq 0$ whenever $x \neq 0$? Certainly the condition

(ND)   Each matrix M_Q is non-singular

will guarantee that $P_M(x)$ is zero only when x is. However, since we are only concerned with the behavior of M_Q on the orthant Q, it would in fact be enough to assume the weaker condition that

(WND). The kernel of the matrix M_Q intersects Q only in the point 0.

A matrix is non-singular is and only if its determinant does not vanish. Since M_Q has columns which are either standard basis vectors or columns of M, it is easy to see that the determinant det M_Q is the determinant of a principal submatrix. More specifically, if the coordinates $i_1, i_2, \ldots, i_k$ are the ones which are negative on Q, then

(9) $\qquad$ det $M_Q = $ det $\begin{vmatrix} m_{i_1 i_1} & \cdots & m_{i_1 i_k} \\ \cdot & & \\ \cdot & & \\ \cdot & & \\ m_{i_k i_1} & \cdots & m_{i_k i_k} \end{vmatrix}$

Therefore, we may reformulate condition (ND) as

(ND)'   Each principal submatrix of M is non-singular.

We summarize the basic facts about Π_M in a lemma.

Lemma 2.2.   The map Π_M on S^{n-1} can be associated to any matrix
M satisfying condition (WND), which is implied by condition (ND)'.
The map Π_M then depends continuously on M. Solving problem (1)
is equivalent to inverting the map Π_M in the sense that a pair
x,y of n-vectors satisfy (1) if and only if

(10) $\Pi_M(z\|z\|^{-1}) = q\|q\|^{-1}$

where z = x-y.   ∎

     For aesthetic reasons, in a geometry-oriented article such as
this one, it is desirable to have as coordinate-free a formulation
as is possible. The present situation will not be very coordinate-
free, since the orthant structure on $\mathbb{R}^n$ is compatible with relatively few coordinate systems. However, the system of orthants
is preserved by:

    i)   Permutations of the coordinates (effected by permuta-
         tion matrices).

    ii) Dilations of the coordinates axes (effected by diagon-
         al matrices).

Dilations may be further decomposed into:

    ii-a) Dilations preserving all orthants (effected by posi-
           tive diagonal matrices).

    ii-b) Dilations preserving the set $\{\pm e_i\}_{i=1}^n$ of standard
           basis vectors and their negatives (effected by diagon-
           al matrices with diagonal entries equal to $+1$).

     These transformations generate a group, which we might call
the orthant group. Every element E of the orthant group can be
written uniquely as a product

(11) $E = SD = S|D|(\text{sgn } D)$

where S is a permutation matrix, D is an non-singular diagonal
matrix, $|D|$ is the matrix whose entries are the absolute values of
the entries of D, and sgn D = $|D|^{-1}D$.

We want to describe a collection of maps that are essentially
the same as the maps P_M, as described by (6), but which exhibit
the full symmetry of the orthant group. Let P be the set of
continuous maps T from R^n to R^n such that the restriction of
T to any orthant is (the restriction of) a linear map. Clearly
P is a vector space under addition. Also if A is an n × n
matrix, then the composition AT is in P if T is. By AT we
mean

$$(AT)(x) = A(Tx) .$$

Although we can follow T ε P by a matrix A, we cannot precede
it by an arbitrary linear transformation and stay in the class P,
because in general, the "breaks" in TA, where it is not linear,
will occur not on the faces of the orthants, but interior to them.
However, if E is in the orthant group then TE is again in P.
Hence P is a vector space allowing multiplication on the left
by all square matrices and multiplications on the right by the
orthant group.

We can parametrize elements of P by pairs of matrices L, M.
For, given L,M, we can define

(12) $T_{L,M}(x) = \frac{1}{2}L(x + |x|) + \frac{1}{2}M(x - |x|).$

Comparing with formula (6) we see that the P_M of that formula is
equal to $T_{I,M}$ as defined in (12), where I is the identity matrix.

Lemma 2.3.   Every element of P is represented uniquely in the
form (12). Precisely, if T ε P, then T = $T_{L,M}$ if and only if

(13) $$T_{|(\mathbb{R}^n)^+} = L_{|(\mathbb{R}^n)^+} \qquad T_{|-(\mathbb{R}^n)^+} = M_{|-(\mathbb{R}^n)^+} \quad .$$

If A is any $n \times n$ matrix, and S is a permutation then we have

(14) $$AT_{L,M} = T_{AL,AM} \qquad T_{L,M}S = T_{LS,MS} \quad .$$

Remark: The first statement is this lemma is lemma A.1 in [K-S].

Proof: The formulas (14) are straightforward verifications. Also,
it is clear by inspection of (12) that if $T = T_{L,M}$, then formulas
(13) hold, which implies that T determines L and M. On the
other hand, if L and M are the linear maps which agree with T
on the positive and negative orthants respectively, then we may
form $T_{L,M}$. The difference $T - T_{L,M}$ will be zero on the positive
and negative orthants, and so will send the standard basis vectors
e_i and their negatives to zero. But any orthant is spanned by a
set consisting of certain of the e_i's and the negatives of the
rest. Therefore $T - T_{L,M} = 0$, or $T = T_{L,M}$, and the lemma is
proved.   ∎

    We may define non-degeneracy similarly to above. By abuse of
notation, given $T \varepsilon P$, and an orthant Q, we will let $T_{|Q}$ stand
for the linear transformation whose restriction to Q agrees with
the restriction of T to Q. We will say T is non-degenerate
if it satisfies

(ND)" For every orthant Q, $T_{|Q}$ is a non-singular matrix.

Denote the set of non-degenerate elements of P by NDP. We will
say T is weakly non-degenerate if it satisfies

(WDN)' For every Q, the intersection of Q with the kernel of
        $T_{|Q}$ is the single point 0. In other words T(Q) is a
        proper cone, i.e., contains no full lines.

Denote this set by WNDP. Clearly WNDP is open in P.

For $T \varepsilon WNDP$, we can define a map Π_T on S^{n-1} in direct analogy with formula (8), viz.,

(15) $$\Pi_T(x) = \|T(x)\|^{-1} T(x), \quad x \varepsilon S^{n-1} .$$

If $T = T_{L,M}$, then we will also write

$$\Pi_{T_{L,M}} = \Pi_{L,M} \quad .$$

If $T = T_{L,M} \varepsilon P$, and L is invertible, then by formula (14) we may write

(16) $$T_{L,M} = LT_{I,L^{-1}M} = LP_{L^{-1}M}$$

where P is as in (6). If $T \varepsilon NDP$, then L is invertible by assumption. Every $T \varepsilon NDP$ may be written uniquely in the form

(17) $$T = AP_M$$

where A is a non-singular $n \times n$ matrix and P_M is as in (6), and M satisfies (ND)'.

From this lemma we see that, at least in the non-degenerate case, inverting the maps $T \varepsilon NDP$, or their associated maps Π_T as in (15), is an essentially trivial generalization of problem (1). Hence we will feel free to discuss arbitrary maps in P, not simply the maps P_M.

To close this section, we will discuss how some well-known transformations of M in problem (1) fit into our formulation of (1) in terms of P. The transformations are

(18a)   Conjugating M by a permutation matrix (i.e., a prin-
          cipal of rearrangement of M)

$$S : M \to M' = SMS^{-1}$$

(18b)   Pivots of M : if M is partitioned

$$M = \begin{bmatrix} A & B \\ C & D \end{bmatrix}$$

where A is k × k and D is (n-k) × (n-k), t
then the pivot (c f. [C2],[T]) of M around
A is

$$M' = \begin{bmatrix} A^{-1} & -A^{-1}B \\ CA^{-1} & D-CA^{-1}B \end{bmatrix}$$

The transformations (18a) are easy to understand. In fact, we
may read off directly from formulas (14) that

(19) $ST_{L,M}S^{-1} = T_{SLS^{-1}, SMS^{-1}}$

whence in particular

(20) $P_{SMS^{-1}} = SP_M S^{-1}$.

That is, the problem (1) for SMS^{-1} is just the conjugate by S
of problem (1) for M. The two problems are therefore essentially
equivalent. The particularly simply form of the transformation
laws (19) and (20) result from the fact that S preserves the
positive and negative orthants, in terms of which the coordinates
$T = T_{L,M}$ were defined.

     We may also conjugate elements of P by dilations. For
positive dilations, formulas similar to (19) and (20) result, but
for non-positive dilations, the transformations are complicated
because the positive orthant is not preserved. For example, let
E_k be the diagonal matrix whose first k diagonal entries are -1,
the rest being +1. Then we can compute that

$$
(21) \qquad P_M E_k = \begin{bmatrix} -A & 0 \\ -C & I \end{bmatrix} P_M'
$$

where M' is as in (18b). Thus principal pivoting arises by multiplying P_M on the right by E_k, or some similar diagonal matrix with ± 1 entries, then expressing the result as the product of P_M', followed by a linear map.

3. Degree Theory

Consider the n-1-sphere S^{n-1}, and let

$$
\phi : S^{n-1} \to S^{n-1}
$$

be any continuous map. Algebraic topology allows us to attach to ϕ an integer, the <u>degree</u> of ϕ, written $\deg \phi$. The basic intuition about degree is that it measures the number of times ϕ wraps the sphere around itself. Thus it is a generalization to n-dimensions of the notion of winding number. See Figure 2.

The basic idea of the paper is to apply the ideas of degree theory to derive facts about the maps P_M, for M satisfying condition (WND). Degree theory in its most usual formulation does not apply directly to the maps P_M and T, but rather to the associated maps Π_M and Π_T defined on S^{n-1} by formulas (8) and (15). Thus at some point the standard results about Π_T must get transcribed into results about T. Because of the very simple relation between T and Π_T, the transcription is not difficult, but it must be done. Since we are treating degree theory as a black box in this paper, it seems simplest to just put the transcription in the black box too, and formulate the results we will use so that they apply directly to T. Thus what we present below is not only a summary, but also a slightly modified account of degree theory. For a more detailed and standard treatment of degree theory, see [G-P], [Lf], or [O-R].

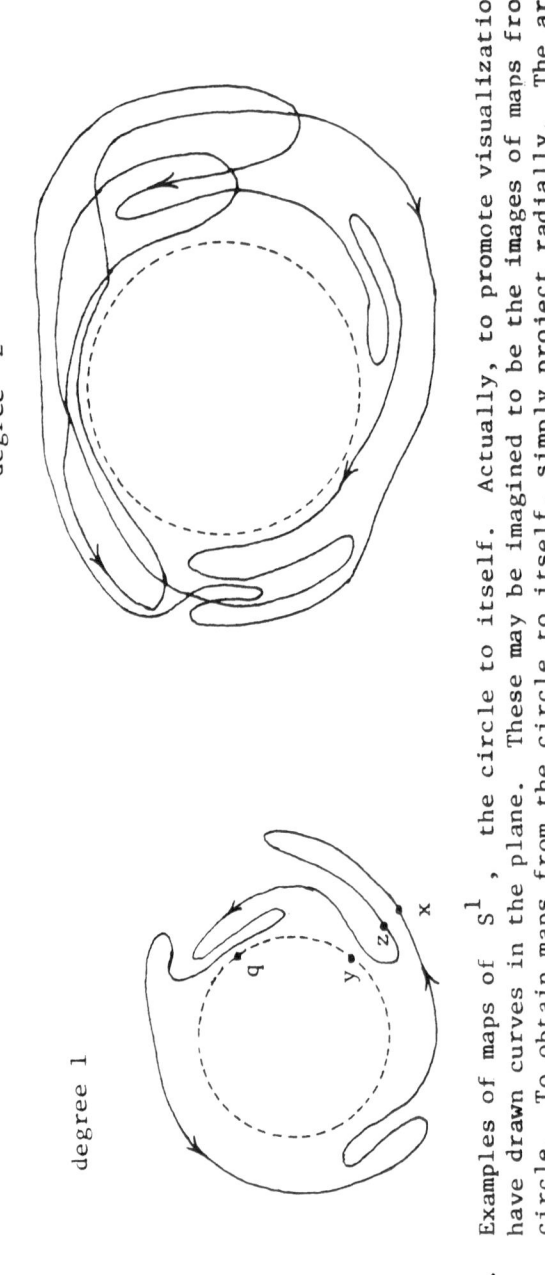

FIGURE 2.   Examples of maps of S^1, the circle to itself. Actually, to promote visualization, we have drawn curves in the plane. These may be imagined to be the images of maps from the circle. To obtain maps from the circle to itself, simply project radially. The arrows → indicate the direction of traversal.

Let $\mu : R^n \to R^n$ be a continuous map. We will say that μ is positive homogeneous of degree 1, or homogeneous for short, if

(22) $$\mu(tx) = t\mu(x), \; x \in R^n, \; t \in R, \; t \geq 0.$$

Clearly, if μ is homogeneous, then $\mu(0) = 0$. We will say μ satisfies (WND) if $x \neq 0$ implies $\mu(x) \neq 0$. That is, if μ satisfies (WND) then μ defines a map from $R^n - \{0\}$ to itself. The sphere S^{n-1} sits inside $R^n - \{0\}$, and there is an obvious projection

(23) $$p : R^n - \{0\} \to S^{n-1}; \; p(x) = \|x\|^{-1}x, \; x \in R^n - \{0\}$$

from $R^n - \{0\}$ onto S^{n-1}. Clearly $p(x) = x$ for x in S^{n-1}. The projection p was implicitly used in defining the maps Π_T. We see that, in fact, given any homogeneous μ satisfying (WND), we can define

$$\Pi_\mu : S^{n-1} \to S^{n-1}$$

by

(24) $$\Pi_\mu(x) = p(\mu(x)), \; x \in S^{n-1}.$$

The usual degree theory would attach an integer deg Π_μ to Π_μ. We will write

(25) $\qquad \deg \mu = \deg \Pi_\mu$.

In the rest of this section, μ will always be a homogeneous map satisfying (WND).

Probably the most basic property of degree is that it is a homotopy invariant. Let μ_0 and μ_1 be two homogeneous maps of R^n to itself satisfying (WND). Recall that the μ's are called homotopic if one can be continuously deformed into the other. Formally, μ_0 and μ_1 are homotopic if there is a continuous map

$$\phi : R^n \times [0,1] \to R^n$$

such that for each t, $0 \le t \le 1$, the map $x \to \phi(x,t)$ of R^n is homogeneous and satisfies (WND), and $\phi(x,0) = \mu_0(x)$ and $\phi(x,1) = \mu_1(x)$ for all $x \in R^n$.

<u>Remark</u>: This is not the notion of homotopy between maps of R^n, but rather the appropriate notion to guarantee the two maps Π_{μ_0} and Π_{μ_1} to be homotopic. The requirement that each $\phi(x,t)$ satisfy (WND) is essential here.

(DEG1) We have deg μ_0 = deg μ_1 if and only if μ_0 and μ_1 are homotopic.

<u>Remark</u>: The "only if" part of (DEG1) is special to spheres.

We will call a homogeneous map μ of R^n a constant map if has for form

$$\mu(x) = t(x)x_0, \quad t(x) > 0 \quad \text{for} \quad x \in R^n - \{0\}$$

for some scalar-valued function t, and some fixed point $x_0 \ne 0$. The terminology derives from the fact that the corresponding map Π_μ of S^{n-1} will take only the value $p(x_0)$. Since for $n > 1$, the sphere S^{n-1} is arcwise connected, any two constant maps are homotopic. More generally, if μ is not surjective, i.e., if not every point of R^n is in the image of μ, then μ is homotopic to a constant map. This is best seen in terms of the squashed map Π_μ, which will certainly be surjective if and only if μ is. If Π_μ omits some point, say the north pole, we may shrink Π_μ to the constant map to the south pole, by pulling it southward along lines of longitude.

(DEG2) A constant map has degree zero. Hence every map of non-zero degree is surjective.

For maps that are smooth or piecewise smooth, such as the elements T of P, there is a beautiful way to express precisely the intuition that a map of degree d covers S^{n-1} exactly d times. This involves the notion of local index. If μ is piecewise smooth, we call a point x in R^n a regular point of μ if i) μ is differentiable at x, and ii) the Jacobian matrix of μ is non-singular. Otherwise, we call x a singular point. Clearly whether x is regular or singular depends only on $p(x) \varepsilon S^{n-1}$, not on where x happens to lie on the ray through $p(x)$. A point $y \varepsilon R^n$ is called a singular value of μ if it is the image under μ of some singular point. The result known as Sard's Theorem tells us that the set of singular values of μ is closed and of measure zero. (For our maps T this will be com-. pletely obvious.)

If x is a regular point of μ, we define the index of μ at x to be the sign of the determinant of the Jacobian matrix, $J_\mu(x)$, of μ at x. That is

(26) $$\text{ind}_\mu(x) = \begin{cases} +1 & \text{if} \quad \det J_\mu(x) > 0 \\ -1 & \text{if} \quad \det J_\mu(x) < 0 . \end{cases}$$

For $y \varepsilon R^n$, set

$$\mu^{-1}(y) = \{x \varepsilon R^n : \mu(x) = y\} .$$

Observe that $\mu^{-1}(y)$ has the same cardinality as $(\Pi_\mu)^{-1}(p(y))$.

(DEG3) For any regular value y of μ, the cardinality of $\mu^{-1}(y)$ is finite, and we have the formula

$$\deg \mu = \sum_{x \varepsilon \mu^{-1}(y)} \text{ind}_\mu(x) .$$

In particular, the cardinality of $\mu^{-1}(y)$ has the same parity as $\deg \mu$.

The reader may wish to verify the formula of (DEG3) for the maps of Figure 2.

The behavior of degree under composition is another very basic property. Suppose μ_1 is another homogeneous map of $\mathbb{R}^n$ satisfying (WND). It is quickly verified that the composed map $\mu_1 \circ \mu$ is again homogeneous and satisfies (WND). Also $\Pi_{\mu_1 \circ \mu} = \Pi_{\mu_1} \circ \Pi_\mu$.

One has

(DEG4) $\deg(\mu_1 \circ \mu) = (\deg \mu_1)(\deg \mu)$.

Finally, especially for use in Section 5, we consider the degree of direct sums. Let μ be as usual and let ν be a homogeneous map of $\mathbb{R}^m$ satisfying (WND). We can define

$$\mu + \nu: \mathbb{R}^{n+m} \to \mathbb{R}^{n+m}$$

by the obvious formula:

(27) $(\mu + \nu)(x, x') = (\mu(x), \nu(x'))$, $x \in \mathbb{R}^n$, $x' \in \mathbb{R}^m$.

It is obvious that $\mu + \nu$ is homogeneous and satisfies (WND).

(DEG5) $\deg(\mu + \nu) = (\deg \mu)(\deg \nu)$.

If μ and ν are piecewise smooth, it is not hard to deduce (DEG5) from the index formula (DEG3).

4. Applications of Degree Theory to Linear Complementarity

In this section, we apply degree theory to the maps T in WNDP, and in particular to the maps P_M of formula (6). Several results are virtually immediate. From (DEG1) and (DEG2), we can make the following conclusions:

Theorem 4.1.   For $T \in$ WNDP, the integer deg T is well defined, and is constant on connected components of WNDP. If deg $T \neq 0$, then T is surjective. In particular, for an $n \times n$ matrix M, if deg $P_M \neq 0$, then M is a Q matrix.

Proof: The only thing that perhaps should be remarked is that an arc T_t, $0 \leq t \leq 1$, between T_0 and T_1 in WNDP defines a homotopy Φ between T_0 and T_1, by simply taking $\Phi(x,t) = T_t(x)$. ∎

Two elements in the same component of (WNDP) will be called linearly homotopic.

Next we note that T is certainly piecewise smooth, so that (DEG3) applies. The second theorem gives some initial conclusions from (DEG3).

Theorem 4.2. For $T \in$ WNDP, and for y a regular value of T, the number of solutions of $T(x) = y$ has the same parity as deg T. In particular, for a matrix M satisfying (WND) if q is a regular value of P_M, the number of solutions of (1) has the same parity as deg P_M. Thus if (1) has an odd number of solutions for some open set of q's, then deg P_M is odd, and M is a Q-matrix. If M is not a Q-matrix, then (1) has an even number of solutions for all q's which are regular values of P_M. ∎

Of course, it is quite possible for deg P_M to be even, in which case M would be a Q-matrix, but (1) would always have an even number of solutions. We will see examples in Section 5.

Another very simple consequence of (DEG3) and (DEG4), when combined with formula (16), reduces the computation of deg T, at least for $T \in$ NDP, to be computation of deg P_M.

Proposition 4.3. a) For an invertible $n \times n$ matrix A, we have

$$(28) \quad \deg A = \text{sign}(\det A) = \begin{cases} +1 & \text{if } \det A > 0 \\ -1 & \text{if } \det A < 0. \end{cases}$$

b) For $T \in$ WNDP, and E in the orthant group,

we have,

(29) $\deg(ATE) = (\deg A)(\deg T)(\deg E) = \pm\deg T.$

Proof: Formula (29) is immediate from formula (28) and (DEG4).
Formula (28) is immediate from the formula of (DEG3), since
$A^{-1}(y)$ for invertible A always consists of just one point.   ∎

   A class of matrices that has figured prominently in the literature on linear complementarity is the class of strictly semimonotone matrices denoted L_* by Eaves [E1] and E by Lemke
[L1] and Cottle [C]. This class is describable in various ways.
In fact Cottle [C] gives several equivalent characterizations of
strict semimonotone matrices; one is that if M is semimonotone,
then problem (1) has a unique solution if q is non-negative.
(This unique solution will clearly be q itself.) Again using
(DEG3) we find

Proposition 4.4.   If T ∈ (WNDP), and $T^{-1}(y)$ consists of a single
point for y in an open set, or for y a regular value of T,
then deg T = ±1. In particular, if M is strictly semimonotone,
then deg P_M = 1, and M is a Q-matrix.

Proof: The first statement is immediate from the formula of (DEG3)
(and the fact that the critical values have no interior). The
sign of deg P_M for M semimonotone comes from noting that the
Jacobian of P_M in the positive orthant is the identity.

                                                                                ∎

Remark.   The same argument shows Eaves' class L_1 of semimonotone
matrices give degree 1 maps as do Garcia's E*(d), d > 0 (when
they are intersected with (WND), so that degree is defined).

So far we have required only very crude properties of the maps
T. We now look more closely at them. First, consider the regular
points of T. It is clear that the interior points of orthant Q
are regular points if and only if $T_{|Q}$ is non-singular. Typically,
this will exhaust the set of regular points; however, it will
occasionally happen that $T_{|Q_1} = T_{|Q_2}$ for two different orthants,
and then some points on the supporting hyperplanes may be regular
also. In any case, we see from the definition of index that for
x in the interior of Q we have

(30) $\text{ind}_T(x) = \text{sign det } T_{|Q}$.

Consider the special case when $T = P_M$ for some n × n matrix M.
Let $\min_Q(M)$ denote the principal minor of M on the right hand
side of formula (9). Then formula (9) tells us that for x in
the interior of Q, we have

(31) $\text{ind}_{P_M}(x) = \text{sign det}(\min_Q(M))$.

This formula (31) reveals the significance of the signs of the
determinants of the minors for the understanding of problem (1).
For $T \in NDP$, we will call the assignment of ±1 to the orthants
according to formula (30) the <u>sign pattern</u> of T. It is clear
that the sign pattern of T will strongly influence the properties
of T. It might be hoped that the sign pattern of T would
determine the degree of T. However, this is not always so, as
is already seen in the 2-dimensional case. In Figure 3, which
lists the sign patterns and degrees of the maps of Figure 1, we
see that maps a) and c) have the same sign patterns but different
degrees.

Nevertheless, some sign patterns do determine the degree of
any T having them, or very nearly do so. The prime example of
this in the literature (see [M], [E1], [S-Th-W]) is that if the

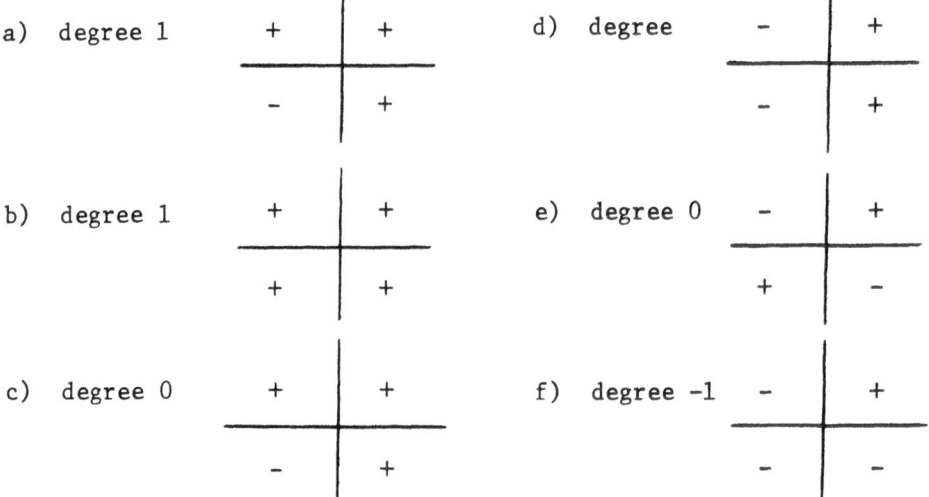

FIGURE 3.   The sign patterns and degrees of the maps of Figure 1.
The degree of P_M in each orthant is indicated by a
+ or or − sign in the orthant.

sign pattern of T consists only of +1's (or if $T = P_M$, then
all principal minors of M have positive determinant, so M is
a so-called P-matrix) then deg T = 1; and moreover T is one-to-
one (so if $T = P_M$, then (1) has always a unique solution). Another result of this sort is the result of Kojima and Saigal [K-S]
who consider the case when all minors of M have negative determinant. We will see how these results can be understood and extended using degree theory.

First we consider the case of P_M with all minors of M
having positive determinant. Whenever all minors of M are non-
zero, we know from Section 2 that $P_M \varepsilon$ NDP. Of course NDP is
a proper open subset of WND , so the connected components of ND
are contained inside connected components of WNDP. We have already agreed to call two elements in the same component of WNDP
linearly homotopic. We will call two elements in the same component of ND strongly linearly homotopic. Clearly two strongly
linearly homotopic elements will have the same sign patterns, but
as the examples of Figure 3 show, the converse is not true.
Similarly, those examples (e.g., a) and b), or c), d), and e))
show that elements with differing sign patterns may be linearly
homotopic. In the other direction, we have the following result.

Theorem 4.5.   Let M be an n × n matrix satisfying (ND)', so
that $P_M \varepsilon$ NDP. Let E be a diagonal matrix with diagonal entries
of ±1. Then P_M is strongly linearly homotopic to P_E if and
only if M has the same sign pattern as E. In particular P_M
is strongly linearly homotopic to I, the identity map, if and
only if all principal minors of M have positive determinant.
In this case the degree of P_M is 1. If M is strongly linearly homotopic to an E ≠ I, then deg P_M = 0.

Remark.   In the examples of Figure 3, example b) has the sign

pattern of $\begin{bmatrix} 1 & 0 \\ 0 & 1 \end{bmatrix}$, example d) the pattern of $\begin{bmatrix} -1 & 0 \\ 0 & 1 \end{bmatrix}$, and e)

of $\begin{bmatrix} -1 & 0 \\ 0 & -1 \end{bmatrix}$. Thus Theorem 4.5 predicts the degree of these maps.

Slightly more abstractly, we see that of the 8 possible 2-dimen-

sional sign patterns with + in the positive orthant (as will

always be the case with P_M), 4 or half of them are accounted for

by diagonal matrices. However with increasing dimension, the

number of sign patterns of diagonal matrices becomes insignificant

compared with the total number of conceivable sign patterns:

2^n out of $2^{(2^n-1)}$.

Before proving Theorem 4.5, we note that it implies the

uniqueness result of [S-Th-W], [M], etc.

Corollary 4.6. ([S-Th-W], [M], etc.)   If M has principal minors

of positive determinant, then there is only one solution of (1)

for all regular values q of P_M.

Proof: By Theorem 4.5, deg P_M = 1. Plug this into the index for-

mula of (DEG3), and observe there can be no cancellation on the

right hand side because all local indices are +1.

                                                                    ∎

Remark.   Actually, it is known that for M with positive principal

minors, system (1) has a unique solution for any q. This slight-

ly more delicate result will follow from the local analysis of

P_M to be given below.

Proof of Theorem 4.5.   Define for $0 \le t \le 1$ the matrix

$$M_t = tE + (1-t)M .$$

Clearly if M_t satisfies (ND)' for all t, then P_{M_t} defines a path in NDP from P_M to P_E. Since E is diagonal, we have for all orthants Q

$$\det \min_Q (M_t) = (\det \min_Q E)(\det \min_Q (t + (1-t)E^{-1}(M)))$$

$$= \pm(\sum_{k=0}^{\ell} t^{\ell-k}(1-t)^k \alpha_k)$$

where ℓ is the rank of $\min_Q M_t$, and α_k is the sum of the determinants of the principal minors of rank k of $\min_Q E^{-1}M$. But since E and M have the same sign pattern, and E is diagonal, the matrix $E^{-1}M$ has all principal minors positive. Hence $\det \min_Q (M_t)$ is a sum of terms all of the same sign, and thus never vanishes. Thus proves the main assertion of the theorem. As for the degrees, it is clear that $\deg P_I = \deg I = 1$. If E has some negative entries, though, we see that P_E cannot be surjective--no vectors with negative coordinates in any place where E has a -1 can be hit by P_E. Hence by Theorem 4.1, we have $\deg P_E = 0$.   ∎

We can increase the applicability of Theorem 4.5 by analyzing the "local structure" of the maps P_M, and we will do this short-ly. First, however, we note some other instances of strong linear homotopy. It is known and easy to show that if D is a positive diagonal matrix, then M is a Q-matrix if and only if DM, or MD, is. Here is an analogue of that result.

Proposition 4.7. If $T = T_{L,M} \in$ NDP, and if D is a positive diagonal matrix, then the maps

$$T_{DL,M} , \; T_{L,DM} , \; T_{LD,M} , \; T_{L,MD}$$

are all strongly linearly homotopic to T.

Proof: Since the group of positive diagonal matrices is connected, it suffices to show for example, that $T_{LD,M}$ is in (NDP) for all D. But in a given quadrant Q, the matrix of $(T_{LD,M})_Q$ is obtained from T_Q by multiplying certain columns by the corresponding entries of D. Thus $(T_{LD,M})_Q$ is certainly invertible if T_Q is, and the proposition follows.

■

Proposition 4.8.   Suppose M satisfying (ND)' has the block triangular form

$$M = \begin{bmatrix} M' & N \\ 0 & M'' \end{bmatrix}$$

where M' is $k \times k$ and M'' is $(n-k) \times (n-k)$. Then P_M is strongly linearly homotopic to $P_{M'} \oplus P_{M''}$. Thus

(32)   deg P_M = (deg $P_{M'}$)(deg $P_{M''}$)   .

Proof: Formula (32) follows directly from the main conclusion and fact (DEG5). To prove the strong homotopy set

$$M_t = \begin{bmatrix} M' & tN \\ 0 & M'' \end{bmatrix} , \quad 0 \le t \le 1.$$

It is easy to see that all principal minors of M_t will have the form

$$\begin{bmatrix} A & tC \\ 0 & B \end{bmatrix}$$

where A is a principal minor of M' and B is a principal minor of M''. Thus all principal minors have determinants independent of t, and P_{M_t} is a path in NDP connecting P_M with $P_{M_1} \oplus P_{M_2}$.

■

We now turn to the "local analysis" of maps in NDP. By a
seminorthant we mean a set obtained by specifying the sign (i.e.,
≥ 0 or ≤ 0) of some of the coordinates. The semiorthant determined by specifying the sign of only one coordinate is a halfspace;
and specifying signs for all the coordinates yields the orthants,
which are the minimal semiorthants.

Let V_1 be the subspace of $\mathbb{R}^n$ determined by requiring certain coordinates to be zero, and the rest to be arbitrary. Let
V_2 be the orthogonal subspace, so that $\mathbb{R}^n = V_1 \oplus V_2$. If Q_1 is
an orthant in V_1, then $S = Q_1 \oplus V_2$ is a semiorthant, and all
semiorthant can be put in this form in a unique way. We call
the decomposition of S into $Q_1 + V_2$ the standard decomposition
of S. We call V_2 the spine of the semiorthant S.

Given $T \in NDP$, we want to study the restriction of T to
the semiorthant S. Let $S = Q_1 + V_2$ be the standard decomposition of S. It will simplify matters if we assume that T restricted to Q_1 is the identity map. We can arrange this by
multiplying T on the left by a linear transformation, if necessary. Also, by permuting the coordinates by means of the orthant
group, we may as well assume that V_1 is defined by setting the
last $n-k$ coordinates equal to zero, and V_2 by setting the
first k coordinates equal to zero. The following arguments are
carried out under these assumptions.

Consider an orthant $Q \subseteq S$, and write $Q = Q_1 \oplus Q_2$ where
Q_2 is an orthant of V_2. We may write T_Q in the form

(33)
$$T_Q = \begin{bmatrix} I_k & N_Q \\ 0 & T_{Q_2} \end{bmatrix}$$

where I_k is the $k \times k$ identity matrix, and T_{Q_2} is an

$(n-k) \times (n-k)$ matrix. Since T is in NDP, the matrix T_Q is

non-singular. Hence the matrix T_{Q_2} is non-singular also. It is

clear that the maps T_{Q_2} fit together to define a continuous map

T_2 from V_2 to itself. But then we see that in fact $T_2 \, \varepsilon \, P_2$,

the analogue of P for V_2; and since the T_{Q_2} are non-singular,

we even have $T_2 \, \varepsilon \, \text{ND}P_2$. We call T_2 the <u>local map around</u> Q_1

determined by T, and we call $\deg T_2$ the <u>local degree</u> of T at

Q_1.

<u>Remark.</u>   Actually, T_2 is intrinsically defined as a map from

the quotient space $\mathbb{R}^n/V_1$ to the quotient space $\mathbb{R}^n/M_Q(V_1)$, where

Q is any orthant in S. Putting T_Q hence T_{Q_2} in the form (33)

requires some choices, and so the form (33) is not canonical.

However, this is irrelevant for our purposes.

Thus, elements of NDP yield hereditarily elements in ND of

spines of seimorthants. Note that elements of WNDP do not have

this property. For if in the discussion above T were only ass-

umed to be in WND , the map T_2 might map some non-zero vectors

to zero and thus fail to satisfy (WND). We remark also that if

$T = P_M$ for some matrix M, and if Q_1 is some face of the posi-

tive orthant, then the local map around Q_1 is P_{M_2}, where M_2

is the obvious principal minor of M.

Another way of looking at the semiorthant $S = Q_1 + V_2$ is as

the union of all orthants containing Q_1. Thus if x is any

point in the relative interior or Q_1, the smallest union of

orthants containing a full neighborhood of x in $\mathbb{R}^n$ is precise-

ly the semiorthant S. A clearly interesting question about the
behavior of T near x is whether T(S) fills up a neighborhood
of T(x) or whether T "folds S over" somehow and x is left
exposed on the edge of T(S).

__Lemma 4.8.__   Let S be a semiorthant and let $S = Q_1 + V_2$ be the
standard decomposition of S. Assume T and S to be normalized
as in equation (33). Let x be a point in the relative interior
of Q_1, and let y be a point in V_2. Take $T \varepsilon NDP$, and let
$T_2 \varepsilon NDP_2$ be the local map around Q_1 determined by T. Then
for all sufficiently small scalars $s > 0$, the cardinalities of
the sets $T^{-1}(x+sy) \cap S$ and $T_2^{-1}(y)$ are the same. In general
$T^{-1}(x+sy) \cap S$ has fewer elements than $T_2^{-1}(y)$. Thus T(S) covers
a neighborhood of T(x) if and only if T_2 is surjective from
V_2 to V_2. Hence if $\deg T_2 \neq 0$, the image of S under T will
cover a neighborhood of x.

__Proof:__   For an orthant $Q \subseteq S$, the matrix T_Q has the form (33).
Suppose $u \varepsilon Q$ and $T(u) = x+sy$. Write $u = u_1 + su_2$ with
$u_1 \varepsilon Q_1$ and $u_2 \varepsilon Q_2$. Below, we consider u_1 and u_2 as n-vectors,
or as dim V_1 and dim V_2 vectors as convenient. No confusion
should arise. Then we see that $T_{Q_2}(u_2) = y$, and $u_1 + sNu_2 = x$.
Hence $u_2 = T_{Q_2}^{-1}(sy)$ is uniquely determined, and so is
$u_1 = x - Nu_2 = x - sNT_{Q_2}^{-1}(y)$. Conversely, suppose $y = T_{Q_2}(u_2)$
for some $u_2 \varepsilon Q_2$. Since x is in the relative interior of Q_1,
the vector $u_1 = x - sNu_2$ will be in Q_1 for all sufficiently
small s, so we can reverse the process and find $u = u_1 + su_2$
such that $T(u) = x+sy$. This establishes a bijection between the
two sets in the first statement of the lemma. The rest of the
lemma follows easily.   ■

Our local analysis allows us to prove a substantial refinement on the uniqueness result, Corollary 4.6. Before stating it
we make an observation about the sign pattern of the local maps.
First note that for a matrix A and a map T ε NDP, the sign
pattern of AT is sign det A times the sign pattern. That is,
if det A > 0, then AT and T have the same sign pattern, while
if det A < 0, the sign pattern of AT is just the reverse of
the sign pattern of T, with -1's in place of +1's and vice versa.
We will call this the reverse sign pattern to the original. Since
multiplying T by a linear map is geometrically not a radical
thing to do to T, we would expect that T's with mutually reverse
sign patterns would be similar in many ways.

The orientation of a spine V_2 of the semi-orthant S is
not determined, so the sign pattern of T_2, the local map around
Q_1, is determined only up to reversal. However from (33) we may
immediately assert:

(34)   With appropriate choice of sign, the sign pattern of T_2,
       the local map around Q_1 determined by T, agrees with
       the sign pattern of T restricted to S (when one makes
       the obvious identification between orthants in S and
       orthants in V_2).

Theorem 4.10.   Let T ε NDP be given, and let S $\subseteq \mathbb{R}^n$ be a semiorthant. Suppose the sign pattern of T gives the same sign to
all the orthants in S. Then T is injective on S. That is,
no two points of S are mapped to the same point by T.

Proof: As with Lemma 4.9, the truth of this theorem is unchanged
if T is composed with an invertible matrix, so we may as well
assume T is normalized as in formula (33). Let $S = Q_1 + V_2$
be the standard decomposition of S, and let V_1 the the span of

Q_1. Choose $y \in \mathbb{R}^n$ and write $y = y_1 + y_2$ with $y_i \in V_i$. The argument of Lemma 4.9 shows there is at most one element of $T^{-1}(y) \cap S$ for each element of $T_2^{-1}(y_2)$. Hence to prove the theorem it is enough to show that T_2 is one-to-one. But since, with appropriate normalization, T_2 has all +1's in its sign pattern, this is just the strong version of Corollary 4.6. We may finish its proof as follows. Suppose for 2 distinct points u and v in V_2 we have $T_2(u) = T_2(v)$. Let S_1' and S_2' be the smallest semiorthants in V_2 containing u and v in their respective interiors. Then the sign patterns are all of one sign. By Theorem 4.5, and the remarks on sign patterns just above, these local maps have degree 1, and so are surjective. Hence T_2 maps S_1' and S_2' both onto a neighborhood of $T_2(u) = T_2(v)$. As the local maps are positive homogeneous of degree 1, one sees that T_2 maps any neighborhood of u onto a neighborhood of $T_2(u)$ and similarly for v. Hence, as T_2 is continuous and piecewise linear on finitely many pieces, any point sufficiently near $T_2(u)$ has distinct T_2-inverse images near both u and v. Choosing such a point to be a regular value of T_2 contradicts corollary 4.6. So T_2 must be one-to-one and the Theorem is proved.   ∎

This local version of uniqueness theorem allows us to estimate the degrees of maps with quite varied sign patterns. We will pursue this theme at some length in Section 5. To finish this section we give one example which connects with the existing literature.

Theorem 4.11. Let $T \in \text{ND } P$ be given. Suppose the sign pattern of T assigns the same sign to all the orthants in a halfspace S. Then $|\deg T| \leq 1$. More precisely either $\deg T = 0$, or $|\deg T| = 1$, and the sign of $\deg T$ is the sign of the orthants in S.

<u>Proof</u>: To show this, it suffices to prove, by the index formula

of (DEG3) that there is a point $y \in \mathbb{R}^n$ which is covered by T

either not at all, or only once and with index the same as the

orthants in S. To prove this, it is clearly enough, by Theorem

4.10 to show there is some point not in T(-S) where -S is the

opposite halfspace to S. This is very easy, but it seems a

sufficiently significant fact to state separately, so we put it

into the next lemma, which will complete the proof of the theorem.

■

<u>Lemma 4.12</u>. For any $T \in P$ and any halfspace S, the image T(S)

is not all of $\mathbb{R}^n$.

<u>Proof</u>: Let $S = Q_1 + V_2$ be the standard decomposition of S.

Since S is a halfspace, Q_1 is a single ray, and V_2 is a

hyperplane. Choose $x \in Q_1 - \{0\}$, and consider the point $-T(x)$.

Let Q be any orthant in S. Then either T_Q is singular or non-

singular. If T_Q is non-singular, then $T_Q(Q)$ is a closed point-

ed cone containing $T(x)$, and so there is an open neighboorhood of

$-T(x)$ disjoint from $T_Q(Q)$. By taking an intersection, we can

find a neighboorhood U of $-T(x)$ disjoint from all $T_Q(Q)$ for

Q such that T_Q is non-singular. But if T_Q is singular, the

cone $T_Q(Q)$ is contained in a linear subspace of $\mathbb{R}^n$ of dimen-

sion less than n. So there must be points of U not contained

in these $T_Q(Q)$ either, and these points are not in T(S).

■

<u>Corollary 4.13</u>. (Kojima and Saigal). Given $T \in NDP$, if the

sign pattern of T assigns all orthants the same sign except for

one orthant which receives the opposite sign, then $|\deg T| \leq 1$

In particular, if M is a matrix all of whose principal minors

have negative determinant, then $|\deg P_M| \leq 1$. Moreover, let Q be

the exceptional orthant. If $\deg T = 0$, then the image of T is

contained in T(Q), and $T^{-1}(q)$ consists of 2 points, one in Q, one not, for $q \, \varepsilon \, \text{int} T(Q)$. If $|\deg T| = 1$, then $T^{-1}(q)$ is a

single point for $q \not\in T(Q)$, and contains 3 points if $q \, \varepsilon \, \text{int} T(Q)$.

Proof: The first statement follows directly from Theorem 4.1.
The statement about numbers of inverse images follows for regular

values of T by plugging into the index formula of (DEG3) and

noting that the only possible cancellations comes from Q. For

non-regular values we may argue as in Theorem 4.10. For example,

suppose deg T = 0, and for some point $q \, \varepsilon \, T(Q)$, there are 2

points, u and v in $T^{-1}(q)$, and not in Q. Let S_1 and S_2

be the minimal orthants containing open neighborhoods of u and

v respectively. Just as in Theorem 4.10, we can reach a contra-

diction if we can show at least one of the local degrees of T on

S_1 or S_2 is non-zero. But if $Q \not\in S_1$, then the sign pattern of

T on S_i is all one sign, so the local degree is   1 by Theorem

4.5. But if $Q \, \varepsilon \, S_i$ for i = 1, 2, then we have u and v both

contained in Q, contradicting our original assumption. The

other cases proceed in the same way.   ∎

Remarks.   a)   In the case $T = P_M$, with M having all principal

minors negative, one can easily distinguish the degree 0 and the

degree 1 cases. In fact, if $\deg P_M = 0$, then we say in Corollary

4.12 that in $P_M \subseteq \mathbb{R}^{n+}$. In particular $M(-\mathbb{R}^{n+}) \subseteq \mathbb{R}^{n+}$, which says

M is negative matrix.

        b)   With slight modification, much of the analysis of

Section 4 can be carried through for $T \, \varepsilon \, \text{WND} P$.

5.   Degree Computations

        One of the facts emerging from Section 4 is that most of the

classes of matrices M which have been considered in the liter-

ature on linear complementarity give rise to maps P_M of degree

1. Thus, it seems natural to wonder if in fact $|\deg T| \leq 1$ for

all $T \varepsilon WNDP$, or at least whether $|\deg T|$ is bounded by some
fairly slow growing (say polynomial) function of n (as in $\mathbb{R}^n$).
In this section we will see that this is definitely not the case;
contrariwise, the possible range of $|\deg T|$ grows exponentially
with n. We will explicitly construct maps of very large degree.
In the other direction, we will give some estimates on $\deg T$.

It is obvious from (DEG3) that on $\mathbb{R}^n$ one has $|\deg T| \leq 2^n$,
since no regular value of T can have more than 2^n inverse
images, one per orthant. One can do a little better than that.

Proposition 5.1.   Choose $T \varepsilon NDP$. Let the sign pattern of T
assign -1 to k orthants. Then

(35) $|\deg T| < \min(\frac{3}{2}k, \frac{1}{2}(2^n - k))$, $k \geq 1$.

In particular, all $T \varepsilon ND$ satisfy

(36) $|\deg T| \leq 3 \cdot 2^{n-3} - 1$ for $n \geq 3$.

Proof: If $k = 1$, then we know from Section 4 that $\deg T \leq 1$.
so (35) clearly holds. So we may assume $k > 1$. If Q and Q'
are any two orthants, we can find a halfspace H such that $Q \subset H$,
and $Q' \subseteq -H$ with H and $-H$ respectively containing ℓ and ℓ'
negative orthants for the sign pattern of T, with $\ell \geq 1 \leq \ell'$ and
of course $\ell + \ell' = k$. If $\ell > 1$, we may further break up H into
quarter spaces, each containing some negative orthants. Continuing
in this fashion, we can break $\mathbb{R}^n$ up into semiorthants
$S_1, S_2, \ldots, S_k$, such that the interiors of the S_j are disjoint
from one another, and such that each S_j contains exactly one
negative orthant. (Some S_j may consist of a single negative
orthant.)

The index formula of (DEG3) says that if $\deg T = d > 0$, then each point in $\mathbb{R}^n$ (except for a set of measure zero) must be covered at least d times positively by T. Since a single orthant is mapped to a convex cone, it will cover less than half of the points of $\mathbb{R}^n$ (measured, say, by taking the measure of the intersection of its image cone with the sphere S^{n-1}) of the points of $\mathbb{R}^n$. The estimate $\deg T < (1/2)(2^n - k)$ follows. Further, by combining Corollary 4.13, describing the local map defined by T on one of the S_j, and Lemma 4.9, we see that $T(S_j)$ can at most cover all points one time positively and a cone of points another time positively. Altogether, this allows for all points being covered less than $(3/2)k$ times positively. The combination of these two estimates yields (35). Clearly the estimate (35) yields the worst result when both of its bounds are equal, that is, when $(3/2)k = (1/2)(2^n - k)$, or $k = 2^{n-2}$. Plugging that value of k in (35), noting that $\det T$ is an integer, and recalling that if $\deg T = d$, then $\deg AT = -d$ if A is a matrix of negative determinant, we obtain (36).

∎

Remarks.   a) For $n = 2$, estimate (35) gives $|\deg T| \leq 1$, which is clearly sharp. For $n = 3$, estimate (36) gives $|\deg T| \leq 2$, and we will see shortly that this is sharp also. For $n = 4$, estimate (36) gives $|\deg T| \leq 5$, but a more refined analysis shows in fact $|\deg T| \leq 3$. Presumably estimate (36) gets progressively worse with increasing n. The reason for this seems to be that, for $\deg T$ to be large, there need to be substantial numbers of negative orthants around to allow the folding of the positive orthants necessary for wrapping the sphere around itself many times. On the other hand, if there are many negative orthants, they begin interfering with themselves and begin to prevent the folding, or they cause too much folding, resulting in large amounts of cancellation in the index formula.

b) One can also conclude from the argument of Proposition 5.1 that when the sign pattern of T has k (-1)'s in it, there are at most 3k inverse images of any point, extending the Kojima-Saigal result.

We will now give some explicit examples of matrices M with $|\deg P_M| > 1$. These examples generalize one of Murty [M].

Example 5.2.   The matrix

$$M_{k,n} = \begin{bmatrix} -1 & 2 & 2 & 2 \cdots & & & & 2 \\ 2 & -1 & & & & & & \\ & & \ddots & & & & & \\ & & & -1 & 2 & 2 & & \\ 2 & & & 2 & -1 & 2 & & \\ & & & 2 & 2 & 1 & & \\ & & & & & & & 2 \\ 2 & & & & & 2 & & 1 \end{bmatrix}$$

defined by

(37) $m_{ii} = -1$ for $1 \le i \le k$, $m_{ii} = 1$ for $k < i \le n$

$$m_{ij} = 2 \quad \text{for} \quad i \ne j$$

yields a map $P_{M_{k,n}} = T_{k,n}$ of degree

(38) $\deg T_{k,n} = 1-k$.

In particular, $\deg T_{n,n} = 1-n$.

<u>Proof</u>:   It will suffice, by (DEG3), to show that any point q
with all positive coordinates is covered once positively and k
times negatively by $T_{k,n} = T$. Such a point q is clearly covered once positively by $T(\mathbb{R}^{n+})$. Let Q_j be the orthant where all
coordinates are positive except the jth. For $1 \leq j \leq k$, we
see that $M_{k,n}(-q_j e_j)$ has its jth coordinate equal to q_j and
its other coordinates negative. Hence

$$q = M_{k,n}(-q_j e_j) + r = T(-q_j e_j + r)$$

where r is a non-negative vector with j^{th} coordinate zero.
Thus q is covered (negatively) by $T(Q_j)$ if $j \leq k$. Since all
columns of $M_{k,n}$ beyond the kth are positive, $T(Q_j)$ will
consist of vectors with non-positive j^{th} coordinates for $j > k$.

Let N be any principal minor of $M_{k,n}$ or rank at least 2.
Since the off-diagonal entries of N are positive and larger in
absolute value thant the diagonal entries it is easy to see that
any positive linear combination of the rows of N can have at
most 1 coordinate negative. Therefore, if Q is the orthant
corresponding to N, $T(Q)$ cannot cover $q > 0$. Hence altogether
q is covered once positively by $\mathbb{R}^{n+}$, and k times negatively
by the $T(Q_j)$ for $j \leq k$ for a total degree of $1-k$, as asserted.

■

Once we have produced maps of degree greater than 1, the
flood gates are open, for we can make the degree grow rapidly by
taking direct sums.

<u>Proposition 5.3.</u>   On $\mathbb{R}^n$, there exists elements $T \varepsilon NDP$ such
that $\deg T \geq 2^{(2/5)n - 1}$.

<u>Proof</u>: It suffices to exhibit a T of this index. Write
$n = 5m + \ell$ with $\ell = 0, 1, 2, 3,$ or 4. We will take $T = P_M$ where

$$M = \begin{bmatrix} M_{5,5} & 0 & \cdots & 0 & 0 \\ 0 & M_{5,5} & & & \\ \cdot & \cdot & & \cdot & \\ \cdot & \cdot & & \cdot & \\ \cdot & \cdot & & \cdot & \\ & & & & 0 \\ 0 & & 0 & M_{5,5} & \\ 0 & & & & \widetilde{M}_{\ell} \end{bmatrix}$$

There are m 5×5 blocks $M_{5,5}$, as in Example 5.2, plus one $\ell \times \ell$ block $\widetilde{M}_{\ell}$ at the end. The $\widehat{M}_{\ell}$ block is the identity if $\ell = 0$ or 1, and is $M_{\ell,\ell}$ of Example 5.2 if $\ell = 2, 3$, or 4. According to (DEG5) and formula (38) we have

$$\deg P_M = 4^m \max(1,\ \ell-1) = 2^{2m} \max(1,\ \ell-1) \geq 2^{(2n/5)-1}$$

as claimed.   ∎

By a continuity argument, one can see that if there is a $T \varepsilon ND^p$ of degree $d > 0$, there is another T of any positive degree less than d. Thus the possibilities of deg T grow exponentially with n. This would seem to make it unlikely that there is any simple method of computing deg T in all cases.

6. <u>Remarks on Algorithms</u>

From the geometric viewpoint adopted here, two classes of algorithms occur as candidates for solving problem (1): path-following algorithms and homotopy algorithms. (Both kinds of algorithms of course involve path-following; but those we call homotopy algorithms involve changing the mapping one wished to invert, whereas those we refer to as "path-following" simply involve inverting a fixed map along a path in the range of the map.) In the former type, one would start with points x and y for which one knew $P_M(x) = y$. (Taking $x = y$ in R^{n+} is an obvious choice.) Then one would draw some path from y to q. (There are many possibilities, the straight line being the obvious first choice.)

One then would attempt to "lift" this path via P_M^{-1}, and follow it

along until one arrives at $P_M^{-1}(q)$. The path following algorithms

with the obvious choices of starting point and path suggested above

essentially amount to the Lemke-type algorithms of [L], [E1], and

[G]. It seems that the algorithm of van der Heyden [vH] is another

path following algorithm, but following a broken line from y to

q with individual segments parellel to the coordinate axes.

The only problem that path-following algorithms could run

into is the following. Let γ be the path from y to q, and

let $\tilde{\gamma}$ be the lift of γ via P_M^{-1}, beginning at x. As we pro-

ceed along $\tilde{\gamma}$, we might come to the interface between two orthants

which have opposed signs in the sign pattern of P_M. When that

happens, proceeding forward along $\tilde{\gamma}$ corresponds via P_M to re-

versing direction and proceeding backwards along γ. Thus it

might happen that after proceeding along $\tilde{\gamma}$ for a while we might

return to $P_M^{-1}(y)$ instead of arriving at $P_M^{-1}(q)$. Thus for ex-

ample in Figure 2 if we start at x and head for the inverse

image of q, we will not reach it, but will return to z lying

over y again. Of course we can persevere, and in the 1-dimension-

al case we eventually will reach a point over q. But in higher

dimensions, one can be caught on a closed loop that will cycle

and cycle and never find q.

One way to be assured that this problem will not occur is to

to start a point y such that $P_M^{-1}(y)$ consists of a single point

x. Thus we see the significance of the Lemke class L_1, for which

all strictly positive q permit only one solution to (1) and the

more refined Garcia classes [G] which for some positive q (hence

all nearby q) there is only one solution to (1). For these

classes path-following techniques will work. However, assuming a

single inverse for a regular value of P_M implies $|\deg P_M| = 1$,

so to guarantee success of the path following technique one must

restrict oneself to maps of degree ± 1. This explains our empirical observation in Section 4 that most of the classes of matrices
considered in connection with (1) did give maps of degree 1.

The homotopy algorithms are not so transparently constructed,
and much less prevalent in the literature. In fact the paper of
Eaves and Scarf [E-S] proposes the only homotopy algorithm known
to the author. The basic idea is to construct a homotopy from the
given P_M to a well-understood map P_{M_0} , solve the problem (1)
for M. This could also be looked on as a path-following technique,
but the path is a path of maps rather than of points. Unlike the
regular path-following technique, the homotopy method poses some
conceptual problems at the outset, namely what standard model M_0
to choose, and how to define the homotopy. About the only map
that immediately strikes one as "well-understood" is the identity
map P_I, or some reasonably mild perturbation of it. But choosing
$I = M_0$ immediately limits one to maps of degree 1 since homotopy
preserves degree. Furthermore, choosing M_0 still leaves one
with a second non-obvious problem, namely how to perform the homotopy.

Despite these difficulties, the homotopy method is **essentially**
of the same power as the path-following method. That is, for the
class of maps for which a path-following technique will certainly
work, namely those for which some regular value is assumed only
once, one can specify a homotopy to a standard map. We describe
how to do this.

Without essential loss of generality, let us assume that for
the vector $q_0 = (1, 1, \ldots, 1)$ there is only the one solution
$x = q_0$, $y = 0$ to (1). Instead of P_I, we will use as our basic
mapping P_{M_0} , where

$$
M_0 = \begin{bmatrix} 1 & \cdots & 1 \\ \vdots & & \\ 1 & & 1 \end{bmatrix}
$$

The reader may convince himself that, indeed P_{M_0} is relatively easy to understand, and $P_{M_0}^{-1}(q)$ consists of a single point for almost all q. For a homotopy we will use the naive one P_{M_t} where

(39) $M_t = (1-t)M_0 + tM, \quad 0 \leq t \leq 1$.

The salient observation then is

<u>Lemma 6.1.</u>   Suppose $P_M \varepsilon$ WNDP, and for $q = q_0$, the system (1) has only the solution $x = q_0$, $y = 0$.   Then M_t, as in (39) is in WND   for all t in $[0,1]$, so that P_{M_t} does constitute a homotopy in WND   from P_M to P_{M_0} .

<u>Proof</u>:   We must show that M_t satisfies the condition (WND) of Section 2 for all t.   Since q_0 and M_0 are invariant under permutations of coordinates, it will be enough to check (WND) for quadrants Q_k, in which the first k coordinates are positive and the rest are negative.   For Q_k, we have (using the notation of (WND))

$$
(M_t)_{Q_k} = \left[\begin{array}{c|c|c|c} I_k & & & \\ \hline & (M_t)_{k+1} & \cdots & (M_t)_n \\ \hline 0 & & & \end{array} \right]
$$

where I_k is the $k \times k$ identity matrix and $(M_t)_j$ is the jth column of M_t. To verify (WND) for Q_k, we must check that the dependence relation

(40)
$$\sum_{i=1}^{k} a_i e_i = \sum_{j=k+1}^{n} b_j (M_t)_j$$

has no solution with non-negative a's and b's not all zero. Suppose (40) does have a solution. Write $(M_t)_j = (1-t)q_0 + tM_j$. Then (40) becomes

(41)
$$\sum_{i=1}^{k} a_i e_i = \sum_{j=k+1}^{n} b_j (1-t)q_0 + \sum_{j=k+1}^{n} b_j tM_j ,$$

or

(42)
$$\sum_{i=1}^{k} a_i e_i - \sum_{j=k+1}^{n} b_j tM_j = (\sum_{j=k+1}^{n} b_j (1-t))q_0 .$$

But (42) just says that the vectors $x = (a_1, \ldots, a_k, 0, \ldots, 0)$ and $y = (0, 0, \ldots, 0, tb_{k+1}, \ldots, tb_n)$ are non-trivial solution to (1) for $q = sq_0$, where $s = (\sum_{j=k+1}^{n} b_j (1-t)) > 0$. But this is impossible by our assumption on M. Hence (40) has no solution and M_t is in (WND) for all t in [0,1]. This finishes the lemma. ∎

Thus we see that all algorithms proposed so far are more or less equally capable of solving problem (1), and that they can do so under the condition that some positive vector have a unique inverse image under P_M, a condition that entails that deg P_M = 1. We will close the paper with some questions that suggested by this conclusion and the other results of the paper.

1. Is it possible to reasonably characterize the class $P_1 = (T \varepsilon \text{ WND}P : \deg T = 1)$?

2. In particular, is P_1 connected, so that every element of P_1 can be deformed in WNDP to the identity map?

3. Are there any naturally occurring problems of type (1) for which $|\deg P_M| > 1$? If so, what can be done about the higher degree case?

One might also wonder if degree was a sufficiently powerful invariant that the converse of Theorem 4.1 were valid. In a similar vein, one might hope that when deg T = 1, there is always some regular value of T with one inverse image, so the known algorithms would apply to T. Unfortunately, the LCP is too complicated for these pleasant thoughts to be true. A counterexample to the converse of Theorem 4.1 was given in [K-W] (see [G-G] for a discussion). In [H], counterexamples to both possibilities are given.

REFERENCES

[C]   R. Cottle, Completely-Q Matrices, Stanford University
           Department of Operations Research Technical Report 79-12,
           Sept. 1979.

[C-2]   R. Cottle, Manifestations of the Schur Complement, Lin.
           Alg. and App. 8, (1974), 189-211.

[C-D]   R. Cottle and G. Dantzig, Complementary Pivot Theory of
           Mathematical Programming, Lin. Alg. and its Applications
           1 (1968), pp. 103-125.

[C-R-S]   R. Cottle, R. von Randow, and R. Stone, On Spherically
           Convex Sets and Q-matrices, Lin. Alg. and App., to
           appear.

[D-L]   R. Doverspike and C. Lemke, A Partial Characterization
           of a Class of Matrices Defined by Solutions of the Linear
           Complementarity Problem, Math. of O. R., to appear.

[E]   B. C. Eaves, The Linear Complementarity Problem, Manage-
           ment Sci. 17 (1971), pp. 612-634.

[E-S]   B. C. Eaves and H. Scarf, The Solution of Systems of
           Piecewise Linear Equations, Math. Op. Res. 1, No. 1
           (1976), pp. 1-27.

[G]   C. Garcia, Some Classes of Matrices in Linear Complement-
           arity Theory, Math. Programming 5, No. 3 (1973), pp. 299-
           310.

[G-G]   C. Garcia and F. Gould, Studies in Linear Complementarity, Center for Math. Studies in Business and Economics, Technical Report 8042, University of Chicago, Nov. 1980.

[G-P]   V. Guillemin and A. Pollack, Differential Topology, Prentice Hall (1974), Engelwood Cliffs, N.J.

[K-W]   L. Kelly and L. Watson, Q-matrices and Spherical Geometry, Lin. Alg. and App. 25 (1979), 175-189.

[vH]   L. van der Heyden, A Variable Dimension Algorithm for the Linear Complementarity Problem, J. F. Kennedy School Discussion Paper Series, Number 67D, June 1979.

[H]   R. Howe, On a class of linear complementarity problems of variable degree, in these Proceedings.

[K]   S. Karamardian, The Complementarity Problems, Math. Programming 2 (1972), pp. 107-129.

[O-R]   J. Ortega and W. Rheinboldt, Iterative Solutions of Non-linear Equations in Several Variables, Academic Press, N.Y. (1970).

[M]   K. G. Murty, On the Number of Solutions to Complementarity Problems and the Spanning Properties of Complementary Cones, Lin. Alg. and Applic. 5 (1972), pp. 65-108.

[K-S]   M. Kojima and R. Saigal, On the Number of Solutions to a Class of Linear Complementarity Problems, Math. Programming 17 (1979), pp. 136-139.

[Lf]   S. Lefschetz, Introduction to Topology, Princeton University Press, Princeton, N.J., 1949.

[L1]   C. Lemke, Bimatrix Equilibrium Points and Mathematical Programming, Management Sci. 11, No. 7 (1965), pp. 681-689.

[L2]   C. Lemke, Recent Results on Complementarity Problems, in Non-Linear Programming, Eds. J. Rose, O. Mangasarian, and K. Ritter, Academic Press, New York, 1970.

[Sa1]   R. Saigal, A Characterization of the Constant Parity of the Number of Solutions to the Linear Complementarity Problem, SIAM J. App. Math. 23 (1972), 40-45.

[Sa2]   R. Saigal, On the Class of Complementary Cones and
           Lemke's Algorithm, SIAM J. App. Math. 23 (1972), 46-60.

[S-Th-W]   H. Samuelson, R. M. Thrall, and O. Wesler, A Partition
           Theorem for Euclidean n-space, Proc. Amer. Math. Soc.
           9 (1958), pp. 805-807.

[St]   R. Stone, Ph.D. Thesis. Stanford University, 1981.

[T]   A. Tucker, A Combinatorial Equivalence of Matrices,
           Proc. Symp. App. Math. X, A.M.S. 1960, 129-140.

[W]   L. Watson, A Variational Approach to the Linear Comple-
           mentarity Problem, Ph.D. thesis, University of Michigan,
           1974.

SUB- AND SUPERSOLUTIONS FOR NONLINEAR OPERATORS:

PROBLEMS OF MONOTONE TYPE

Michael Prüfer

Fachbereich Mathematik
Universität Bremen
2800 Bremen 33, W. Germany

0. INTRODUCTION

In this paper we describe for a variety of non-linear problems $F:D \to R^n$, D open in R^n , a discrete programming approach for the calculation of alternating approximations for a zero $z*$:

(0.1)

$$x^1 \preceq \ldots \preceq x^{2k+1} \preceq x^{2k+3} \preceq \ldots \preceq z* \preceq \ldots$$
$$\ldots \preceq x^{2k+4} \preceq x^{2k+2} \preceq \ldots \preceq x^2$$
$$\lim_{k \to \infty} (x^k - x^{k-1}) = 0 ,$$

where $\preceq$, in a general setting, is a partial ordering of R^n . Approximation schemes of type (0.1) are of particular interest if $\preceq$ is the componentwise partial ordering of R^n , denoted $\leq$ throughout this paper: Any two successive elements of the approximating sequence provide componentwise inclusions of $z*$, and hence we have exact error estimates and trivial stopping criteria when calculating $z*$.

In our approach, a sequence (0.1) will alternate between the sets

(0.2)
$$K^- = \{x \in D \mid F(x) \leq 0\} ,$$
$$K^+ = \{x \in D \mid F(x) \geq 0\} .$$

Elements from K^- (K^+) will be called *sub- (super-)*

---

*Work supported by "Forschungsschwerpunkt 'Dynamische Systeme', Universität Bremen"

solutions , a terminology adapted from the theory of
problems of *monotone type* [4] , [5] :
If F has the property that

$$(0.3) \quad F(x) \le F(y) \quad \Rightarrow \quad x \le y \quad \text{for} \quad x, y \in D$$

then we have

$$x \in K_+^- \quad \Rightarrow \quad x \le z^* ,$$
$$x \in K^+ \quad \Rightarrow \quad x \ge z^* ,$$

and a sequence $\{x^k\}$ approximating z^* and alternating
between K^- and K^+ necessarily is of type (0.1)
with $\precsim = \le$.

In [9] a combinatorial search for sub- and supersolutions is described for *M-functions* , problems of
monotone type (0.3) with the additional property

$$(0.4) \quad \begin{aligned} &\text{for any} \quad x = (x_1, \ldots, x_n) \in D \quad \text{the mapping} \\ &\varphi_{i,j}(t) = F_i(x_1, \ldots, x_{j-1}, x_j+t, x_{j+1}, \ldots, x_n) \\ &\text{is monotonically decreasing on its domain} \\ &\text{if} \quad i \ne j \;. \end{aligned}$$

The algorithm [9] generates sub- and supersolutions by suitably lowering (lifting) the components of
a starting vector $x \in D$. This procedure can be viewed
as a discrete combinatorial programming process with
the standard basis $\{e^j\}$ of R^n taken as "search
directions". In [10] it is announced that [9] can
be generalized by replacing the standard basis by a
more general set of search directions. The aim of this
paper is to show that this modification provides a
considerable additional flexibility of the process
(improvement of numerical properties, broader classes
of applications).

I. A SEARCH FOR SUB- AND SUPERSOLUTIONS

In this section we develop the general setup for an
algorithm for sub- and supersolutions. For convenience
of the reader and in order to keep this paper self-con-
tained we give a review of the main results from
[9] , [10] .
Let $C = \{c^1, \ldots, c^n\} \subset R^n$ be a set of linearly independent vectors with the property (cf. (0.4))

$$(1.1) \quad \text{for any} \quad x \in D \quad \text{the function}$$
$$\varphi_{i,j}(t) = F_i(x+tc^j) \quad \text{is}$$

- strictly monotonically increasing, if $i = j$,
- monotonically decreasing, if $i \neq j$.

As indicated in the introduction a set C satisfying (1.1) will play a role of "search directions" in a discrete programming process to be described below. The following trivial lemma will turn out to provide search directions for a variety of problems:

LEMMA 1 : Let $F:D \rightarrow R^n$ be continuously differentiable and let $C = \{c^1, \ldots, c^n\}$ be such that

$$DF(x)_i(c^j) \begin{cases} > 0 , & \text{if } i = j \\ \leq 0 , & \text{if } i \neq j \end{cases}$$

for all $x \in D$. Then C is a set with property (1.1). ●

For M-functions (0.3), (0.4) it is easily verified that $F_i(x_1, \ldots, x_{i-1}, x_i+t, x_{i+1}, \ldots, x_n)$ is strictly monotonically increasing for all $x \in D$, $1 \leq i \leq n$, which means that in this case the standard basis of R^n is always a suitable choice of vectors (1.1).

The set C induces a cone

$$(1.2) \quad K_C = \{x \mid x = \sum_{j=1}^{n} a_j c^j , a_j \geq 0\}$$

and a partial ordering $\leq_C$ on R^n by

$$(1.3) \quad x \leq_C y \Leftrightarrow y - x \in K_C .$$

Note that

$$x \leq_{C_1} y \Rightarrow x \leq_{C_2} y$$

whenever $C_1 \subset C_2$. Therefore

$$x \leq_C y \Rightarrow x \leq y$$

($\leq$ denotes the componentwise partial ordering of R^n), whenever $C \subset R^n_+$.

Some further notation will be useful:

Let $x \in D$, and let $\delta = (\delta_1, \ldots, \delta_n)$, $\delta_i > 0$, be a vector of steplengths. Denote $V(x, \delta, C) =$
$\{x + \sum_{j=1}^{n} a_j \delta_j c^j \mid a_j \in Z\}$, $v^+(x, \delta, C) = \{x + \sum_{j=1}^{n} a_j \delta_j c^j \mid a_j \in Z_+\}$, $v^-(x, \delta, C) = \{x + \sum_{j=1}^{n} a_j \delta_j c^j \mid a_j \in Z_-\}$, where

$Z_+ = \{0, 1, 2, \ldots\}$, $Z_- = -Z_+$. If $x \leq_C y$ we define
the *order interval spanned by* x *and* y by
$[x,y]_C = \{z \mid x \leq_C z \leq_C y\}$.

Starting with $x \in D$ and a vector $\delta = (\delta_1, \ldots, \delta_n)$
of steplengths we want to construct a supersolution
$\bar{x} \in K^+$. Consider the combinatorial algorithm in the
following flow-chart:

(1.4) $\left\{\begin{array}{l} \text{(1.4.1)} \ \underline{\text{Input}}: x \in D \ , \ \delta = (\delta_1, \ldots, \delta_n) \ , \ \delta_i > 0. \\[4pt] \text{(1.4.2)} \ \text{If} \ x \in K^+ \ \text{then stop.} \ \underline{\text{Output}} \ x \ . \\[4pt] \text{(1.4.3)} \ \text{Choose a nonempty set} \ \bar{J} \subset I = \\ \hspace{4.5cm} = \{i \mid F_i(x) < 0\} \ . \\[4pt] \text{(1.4.4)} \ x \leftarrow x + \sum_{j \in \bar{J}} \delta_j c^j \ , \ \text{go to (1.4.2)} \ . \end{array}\right.$

Note that (1.4) generates a sequence in the discrete
set $v^+(x,\delta,C)$ which is monotonically increasing in the
partial ordering $\leq_C$.

THEOREM 1: Let $x \in D$, $\delta = (\delta_1, \ldots, \delta_n)$, $\delta_i > 0$,
and $y \in v^+(x,\delta,C) \cap K^+$. Then the sequence $\{x^k\}$
generated by (1.4) cannot leave the order interval
$[x,y]_C$ and hence must stop with an $\bar{x} \in K_+$.

Proof: Let $y = x + \sum_{j=1}^{n} \alpha_j \delta_j c^j$ with $\alpha_j \in Z$, $\alpha_j \geq 0$.
Assume that for some k we have $x^k \in \partial([x,y]_C)$, i. e.
$x^k = x + \sum_{j=1}^{n} \beta_j \delta_j c^j$, $\beta_j \in Z$, $0 \leq \beta_j \leq \alpha_j$ and $\beta_r = \alpha_r$
for some r . Then by (1.1) we have $F_r(x^k) \geq F_r(y) \geq 0$.
Hence $r \notin I$ in (1.4.3) and this completes the proof
since a step into c^r-direction is forbidden in (1.4.4).●

The element $\bar{x}$ generated by (1.4) clearly is a
minimal element in the set $v^+(x,\delta,C) \cap K^+$ subject
to $\leq_C$. If $\bar{x} \neq \tilde{x}$ is a second minimal element we have
$\bar{x} \not\leq_C \tilde{x}$, $\tilde{x} \not\leq_C \bar{x}$, which means that the sequence $\{x^k\}$
generated by (1.4) must have left the order interval
$[x,\tilde{x}]_C$. This contradicts theorem 1 and we have a unique
minimal element in $v^+(x,\delta,C) \cap K^+$ subject to $\leq_C$.

If we interchange + and - , < and > in (1.4)
(except for $\delta_i > 0$ in the input) we obtain a strategy
with dual properties:

THEOREM 2: Let $x \in D$, $\delta = (\delta_1,\ldots,\delta_n)$, $\delta_i > 0$,
and $y \in V^-(x,\delta,C) \cap K^-$. Then the sequence $\{x^k\}$
generated by (1.4) and modified as stated above
cannot leave the order interval $[y,x]_C$ and hence
must stop with an $\underline{x} \in K^-$. ●

An analogous argumentation as before shows that $\underline{x}$ is
the unique maximal element in $V^-(x,\delta,C) \cap K^-$ subject
to $\leq_C$.

Our argumentation so for yields sub- and supersolutions for z^* provided a set C of search directions (1.1) can be found. The following figure illustrates a search for a zero of an M-function in two
dimensions, where C has been chosen to be the standard
basis and $J = I$ in (1.4.3).

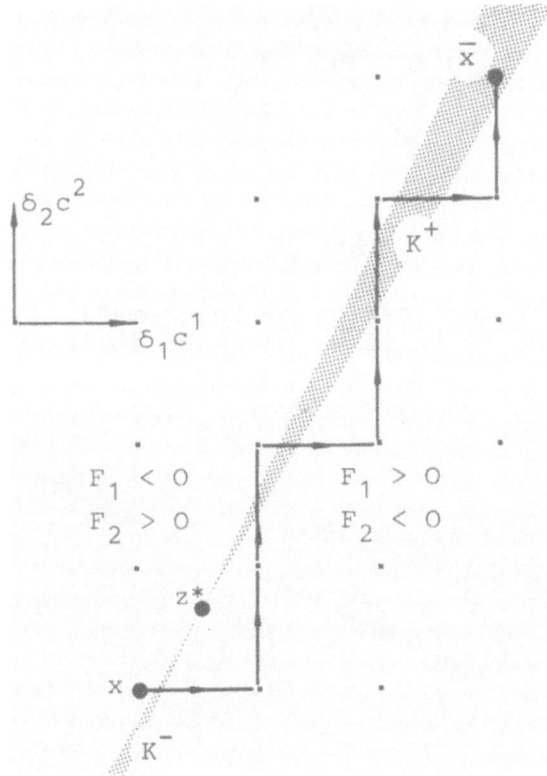

Fig. 1

Figure 1 exhibits a bad numerical situation since, due
to the skinny shape of K^+, $\bar{x}$ is located at a considerable distance from z^*. Figure 2 indicates that one can
hope for better numerical results (less search steps) if
the cone K_C is "adapted" to the set K^+. In the next
section we will discuss how satisfactory search directions can be found for a variety of problems.

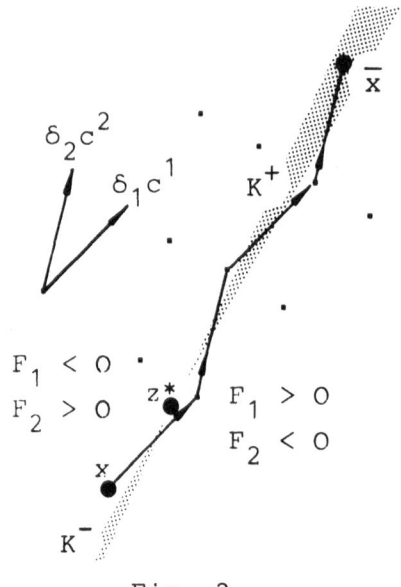

Fig. 2

An alternate application of theorems 1, 2 hopefully
yields a sequence (0.1) if we make the following assumption:

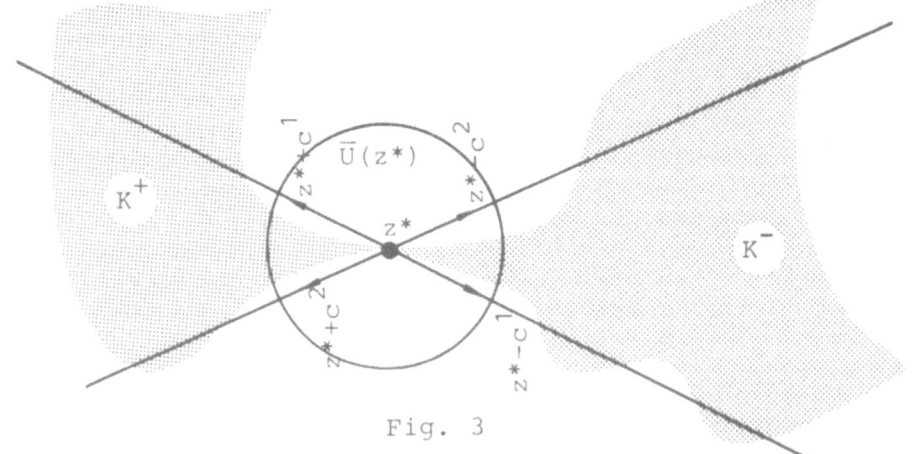

Fig. 3

(1.5) For·any open neighborhood $U(z^*)$ of z^* we
have $U(z^*) \cap \text{int } K^- \neq \emptyset \neq U(z^*) \cap \text{int } K^+$
and there exists an open neighborhood $\bar{U}(z^*)$
such that $\bar{U}(z^*) \cap K^- \subset z^* - K_C$,
$\bar{U}(z^*) \cap K^+ \subset z^* + K_C$.

Let $x^0 \in D$, $\delta^0 = (\delta_1^0,\ldots,\delta_n^0)$, $\delta_i^0 > 0$, and
assume that $V^-(x^0,\delta^0,C) \neq \emptyset$. We use (1.4) modified as
in theorem 2 and obtain by theorem 2 a subsolution
$\underline{x}^1 \in K^- \cap V^-(x^0,\delta^0,C)$. Choose $\delta^1 = (\delta_1^1,\ldots,\delta_n^1)$,
$0 < \delta_i^1 \leq \delta_i^0$. If $V^+(\underline{x}^1,\delta^1,C) \neq \emptyset$ we obtain by (1.4) and
theorem 1 a supersolution $\bar{x}^2$. Note that during the
computation of $\bar{x}^2$ we also obtain an updated subsolution
x^3 : If S is the sequence generated by (1.4) starting
with $\underline{x}^1$ and ending with $\bar{x}^2$ let $\underline{x}^3$ be the maximal
element (subject to $\leq_C$) in $S \cap K^- \cap V^-(\underline{x}^1,\delta^1,C)$. It
is clear that by a repeated application of (1.4) we
obtain a sequence $\{x^k\}$ alternating between K^- and
K^+ , provided the assumptions of theorem 1 and theorem 2
are always fulfilled. If the steplengths involved in
the process tend to zero $(\lim\limits_{k\to\infty} (\max\limits_{1\leq i\leq n} \delta_i^k) = 0)$ and if
$\{x^k\}$ is (at least from a certain index on) contained in
$\bar{U}(z^*)$ (recall (1.5)), then we have an approximation
scheme of type (0.1).
        If z^* is a regular zero of a differentiable
M-function $F:R^n \to R^n$ condition (1.5) holds globally
$(\bar{U}(z^*) = R^n)$, if C is the standard basis of R^n .
Therefore, an arbitrary $x^0 \in R^n$ serves as a starting
point for our process. We remark that a combinatorial
algorithm for the approximation of the zero an M-function
has been described by W.C. Rheinboldt and C.K. Mesztenyi
[12] . Their procedure generates by suitable steps in
single components of an argument vector a sequence in
K^- (or K^+) converging against z^* monotonically from
below (or above). As a starting point [12] explicitly
needs a vector from K^- (or K^+) .

II. OPERATORS OF MONOTONE TYPE AND
ELLIPTIC BOUNDARY VALUE PROBLEMS

M-functions or, more generally, operators of monotone type occasionally arise as discretizations of (nonlinear) ordinary or partial differential equations [3] ,
[4] , [5] , [7] , [11] , [13] . The following theorem
describes how search directions can be found for a class
of such problems. It will be necessary in the sequel to
distinguish by subscripts sub- and supersolutions for
different operators: We write K_F^- , K_G^- , K_F^+ , K_G^+ , etc..

THEOREM 3: Let $F:D \rightarrow R^n$ be continuously differentiable and of monotone type. Assume that there
exists a matrix B of monotone type such that
$DF(x) \leq B$ for all $x \in D$. Let $c^j = B^{-1}(e^j)$,
$1 \leq j \leq n$, ($e_i^j = 0$, if $i \neq j$, $e_j^j = 1$) .
Then the set $C = \{c^1,...,c^n\}$ has property (1.1).

Proof: Since B is of monotone type we have
$B^{-1} \geq 0$, and hence $C \subset R_+^n$. Let $x \in D$ and note that,
since F is of monotone type, $DF(x)$ is of monotone
type. From $DF(x) \leq B$ we conclude that $DF(x)(c^j) \leq$
$\leq Bc^j = e^j$ and $(DF(x))_i(c^j) \leq 0$ for $i \neq j$. But this
implies that $(DF(x))_j(c^j) > 0$ since otherwise
$K_{DF(x)}^- \cap (R_+^n \setminus \{0\}) \neq \emptyset$. The theorem now follows from
lemma 1. ●

Unfortunately, search directions obtained by theorem 3 do not necessarily comply with (1.5). It can be
shown however that they yield reasonable results in view
of (0.1): Let $F:D \rightarrow R^n$ be as in theorem 3 and let $z*$
be a (necessarily unique) zero of F . Then the sets
$z* + K_{DF(z*)}^-$, $z* + K_{DF(z*)}^+$ are *tangential cones* to the
sets K_F^- , K_F^+ . We don't find it necessary to formalize
this largely self-explanatory notion and refer to
figure 4 , instead. Let B and C be as in theorem 3
and note that $K_{DF(z*)}^- \subset -K_C$, $K_{DF(z*)}^+ \subset K_C$, since
$0 \leq DF(z*) \leq B$. Therefore, $z* - K_C$ (resp. $z* + K_C$)
contains the tangential cone of K_F^- (resp. K_F^+) and

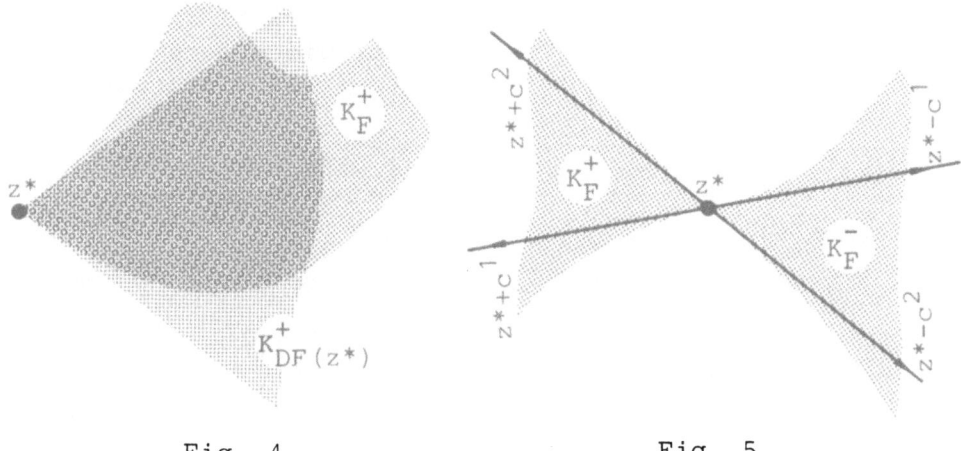

Fig. 4   Fig. 5

figure 5 exhibits a "worst case" situation where our algorithm would provide a "second order" perturbation of a
sequence (0.1).

Example 1: Discretizations of nonlinear elliptic
boundary value problems.

As a typical application we consider the elliptic
two point boundary value problem

(2.1) $\begin{aligned} -u'' + f(u) &= 0 \\ u(0) = u(1) &= 0 \end{aligned}$

discretized on n equidistant interior gridpoints

$$F(x) = Ax + [f](x) = 0 ,$$

$$(2.2) \quad A = \begin{pmatrix} 2 & -1 & 0 & \cdots & 0 \\ -1 & 2 & & & \vdots \\ 0 & & \ddots & & 0 \\ \vdots & & & \ddots & -1 \\ 0 & \cdots & 0 & -1 & 2 \end{pmatrix} , \quad [f](x) = \frac{1}{(n+1)^2} \begin{pmatrix} f(x_1) \\ \vdots \\ \vdots \\ f(x_n) \end{pmatrix} .$$

It is well known that A is an M-matrix and that F is
an M-function if the nonlinearity f is monotonically
increasing [7, 13.5.6] .
        Numerical experience reported in [9] indicates
that our algorithm (1.4) gets very costly as n increases when C is chosen to be the standard basis
of R^n in (1.4.4). The reason is that the sets K^+ and

$\overline{K}$ get thin very rapidly with increasing   n   which puts
us into the unpleasant situation of figure 1   (cf.   [8]
[9, 4.1 and lemma 2]) . Theorem 3 provides more suitable
search directions if we assume that   f   is differentiable
and   f'   bounded:

(2.3) $0 \le f'(t) \le m$ for all $t \in R$.

The matrix

$$(2.4)\quad B = A + \frac{1}{(n+1)^2}\begin{pmatrix} m & 0 & \cdots\cdots & 0 \\ 0 & & & \vdots \\ \vdots & & & \\ \vdots & & & 0 \\ 0 & \cdots\cdots & 0 & m \end{pmatrix}$$

is again an M-matrix   [2, lemma 6.4.1]   and for every

$x \in R^n$ we have $DF(x) \le B$. Therefore, we obtain by
theorem 3 a set C of search directions for our problem.
Applying theorem 3 to problem (2.2) is, of course, motivated by the hope that $z* + K_C$ is a better approximation (locally) to K^+ than the set $z* + R^n_+$ (recall
figures 1 and 2).
   Tables 1 and 2 below reflect computer runs for problem
(2.2) with $f(u) = \exp(u)$, n = 50 . Note that (2.3) is
violated but   f   being positive any solution of (2.1)
must be negative. Therefore, $f\big|_{[0,+\infty)}$ is irrelevant and

m = 1   an appropriate choice in (2.3), (2.4). In table 1

C has been chosen to be the standard basis of R^n ,
whereas in table 2 C has been obtained by theorem 3
and (2.4). In both cases the origin (a supersolution) has
been taken as startingpoint. The vector $\delta^0 = (\delta^0_1,\ldots,\delta^0_n)$
of initial step sizes has been chosen by $\delta^0_i = 0.1$,
and the stepsize has been reduced according to the rule
$\delta^{k+1} = \mu\delta^k$, $\mu = 0.125$. Both computerruns have been performed with   J = I   in (1.4.3).
   In view of theorem 3 an analogous discussion of
problem (2.1) is possible if a higher order approximation
is chosen for the linear differential operator. In a
fourth order approximation [3, section 2] , for instance, the matrix   A   in (2.2) would be replaced by the
matrix

Table 1

k	number of search steps (1.4) from $\bar{x}^0$ to $(\bar{x}^{2k}, \underline{x}^{2k-1})$	$\lVert \bar{x}^{2k} - \underline{x}^{2k-1} \rVert_\infty$	$\lVert F(\bar{x}^{2k}) \rVert_\infty$
1	626	32.5×10^0	1.0×10^{-1}
2	3465	30.8×10^0	3.8×10^{-2}
3	4091	5.1×10^{-1}	1.3×10^{-3}
4	6606	4.4×10^{-1}	1.9×10^{-4}
5	9036	5.4×10^{-2}	2.4×10^{-5}
6	10914	4.9×10^{-3}	3.0×10^{-6}
7	12546	5.2×10^{-4}	3.7×10^{-7}
8	14111	6.4×10^{-5}	6.8×10^{-8}
9	15435	6.3×10^{-6}	6.0×10^{-9}
10	16925	1.1×10^{-6}	7.3×10^{-10}
11	18232	2.8×10^{-7}	1.9×10^{-10}

Table 2

k	number of search steps (1.4) from $\bar{x}^0$ to $(\bar{x}^{2k}, \underline{x}^{2k-1})$	$\lVert \bar{x}^{2k} - \underline{x}^{2k-1} \rVert_\infty$	$\lVert F(\bar{x}^{2k}) \rVert_\infty$
1	11	3.9×10^{-1}	6.0×10^{-3}
2	50	3.2×10^{-1}	6.3×10^{-4}
3	93	4.6×10^{-2}	1.3×10^{-4}
4	134	4.8×10^{-3}	2.4×10^{-5}
5	171	6.0×10^{-4}	2.5×10^{-6}
6	203	6.6×10^{-5}	1.6×10^{-7}
7	244	1.0×10^{-5}	5.2×10^{-8}
8	286	1.3×10^{-6}	3.6×10^{-9}
9	326	1.5×10^{-7}	4.3×10^{-10}

$$(2.5) \; \tilde{A} = \frac{1}{12} \begin{pmatrix} 24 & -12 & 0 & 0 & \cdots\cdots & 0 \\ -16 & 30 & -16 & 1 & & \\ 1 & -16 & 30 & -16 & 1 & \\ 0 & & & & & \\ \vdots & & & & & \\ & & & 1 & -16 & 30 & -16 & 1 \\ & & & & 1 & -16 & 30 & -16 \\ 0 & \cdots\cdots & & & 0 & 0 & -12 & 24 \end{pmatrix}$$

For various discrete approximations (including (2.5)) of -u'' subject to the boundary conditions in (2.1) the monotone type property (0.3) can be proved [3, 6] .

The matrix $\tilde{A}$, however, shows a typical feature of higher order approximations: Positive off-diagonal entries appear and destroy property (0.4). This implies that F (2.2) is no longer an M-function and, in particular, the standard basis of R^n is no longer a set of search directions (1.1) for computing sub- and supersolutions via (1.4). Considerable effort has been paid to the problem: Under what circumstances is the monotone type property of the linear part of a discretized differential equation inherited to a nonlinear perturbation (cf. e. g. [3, section 5]) . For the finite difference scheme (2.2), where f has property (2.3), the corresponding question could be easily clarified, since we had the elementary result [7, 13.5.6] on M-functions. For operators of monotone type in absence of property (0.4), however, theoretical results are not too convenient to handle, whence the following approach seems to be reasonable (as a prototype problem we consider one more time equation (2.1) with f(u) = exp(u)): Let

$$\tilde{B} = \tilde{A} + \frac{1}{(n+1)^2} \begin{pmatrix} 1 & 0 & \cdots & 0 \\ 0 & & & \\ \vdots & & & 0 \\ 0 & \cdots & 0 & 1 \end{pmatrix} .$$

In view of an application of theorem 3 we have to compute $\tilde{B}^{-1}$ anyway and doing so we find $\tilde{B}^{-1} \geq 0$. For any $x \in R^n$ we have $\tilde{A} \leq D\tilde{F}(x) \leq \tilde{B}$, where $\tilde{F} = \tilde{A} + \frac{1}{(n+1)^2}[f]$.

Since $\tilde{A}$ and $\tilde{B}$ are of monotone type $D\tilde{F}(x)$ must be of monotone type [2, (N_{40}) on p. 137] . Hence $\tilde{F}$ is of

of monotone type , and theorem 3 provides search directions (1.1) for sub- and supersolutions for $\widetilde{F}$. The numerical results show but a slight difference to those obtained for F (2.2) in table 2.

Example 2: Operators of monotone type with
isotone derivative.

Let x , y $\in$ R^n , $x_i < y_i$ for all $1 \leq i \leq n$, and
let [x,y] = {z $\in$ R^n | x $\leq$ z $\leq$ y} be the order interval
spanned by x and y . Let F: [x,y] $\rightarrow$ R^n be a continuously differentiable operator such that $F'(z)^{-1}$ is
positive for all z $\in$ [x,y] and $F(x) \leq 0 \leq F(y)$.
Assume furthermore that F' is *isotone* on [x,y] :

$$(2.6) \quad x \leq z_1 \leq z_2 \leq y \;\Rightarrow\; F'(z_1) \leq F'(z_2) .$$

Recall that such a problem is well understood with regard
to sub- and supersolutions, if we have *order-convexity*
as an additional property:

$$(2.7) \quad x \leq z_1 \leq z_2 \leq y , \quad \lambda \in [0,1] \quad \Rightarrow$$
$$F(\lambda z_1 + (1-\lambda)z_2) \leq \lambda F(z_1) + (1-\lambda)F(z_2) .$$

Under these assumptions the *Newton iterates* $y^0 = y$,
$y^{k+1} = y^k - F'(y^k)^{-1}F(y^k)$, can be shown to converge
monotonically (from above) against a unique zero z* .
Minor additional work provides approximations from below:
The sequence $x^0 = x$, $x^{k+1} = x^k - F'(y^k)^{-1}F(x^k)$,
converges against z* monotonically from below [1] ,
[7, thm. 13.3.4] , [14] .
    Note that our approach yields sub- and supersolu-
tions *without* the additional assumption of order con-
vexity: F' being isotone, B = F'(y) fulfills the conditions of theorem 3 and we obtain suitable search directions for our process (1.4). Furthermore, we have the
opportunity to update search directions: Any time a
supersolution $\bar{x}^k \leq y$ has been calculated we may apply
theorem 3 with $B = F'(\bar{x}^k)$.
    Equation (2.1) with f = exp served us as a test
problem and for illustrating the impact of the choice of
search directions on the efficiency of our algorithm. As
a matter of fact, problem (2.2) (and its modification $\widetilde{F}$
with $\widetilde{A}$ (2.5) replacing A) meets the conditions (2.6),
(2.7) and therefore one might want to apply the monotone
Newton method just described. For any problem of this

type, however, our algorithm seems to be an appropriate
tool to find suitable vectors $x \in K^-$, $y \in K^+$ to start
a fast iterative procedure.

REFERENCES

[1] A. Baluev: On the method of Chaplygin (russian),
     Dokl. Akad. Nauk SSSR 83 1952, pp. 781 - 784
[2] A. Berman and R.J. Plemmons: Nonnegative Matrices
     in the Mathematical Sciences, Academic Press,
     New York, San Francisco, London 1979
[3] E. Bohl and J. Lorenz: Inverse monotonicity and
     difference schemes of higher order. A summary for
     two point boundary value problems, Aequationes
     Mathematicae 19 1979, pp. 1 - 36
[4] L. Collatz: Aufgaben monotoner Art, Arch. Math. 3
     1952, pp. 365 - 376
[5] L. Collatz: Funktionalanalysis und Numerische Mathe-
     matik, Springer Verlag, Berlin 1964
[6] J. Lorenz: Zur Inversmonotonie diskreter Probleme,
     Numer. Math. 27 1977, pp. 227 - 238
[7] J.M. Ortega and W.C. Rheinboldt: Iterative Solution
     of Nonlinear Equations in Several Variables,
     Academic Press, New York, San Francisco, London 1970
[8] H.O. Peitgen and M. Prüfer: The Leray-Schauder con-
     tinuation method is a constructive element in the
     numerical study of nonlinear eigenvalue- and bifur-
     cation problems, in "Functional Differential Equa-
     tions and Approximation of Fixed Points",
     H.O. Peitgen and H.O. Walther, eds., Springer
     Lecture Notes in Mathematics 730, Berlin, Heidel-
     berg, New York 1979, pp. 326 - 409
[9] M. Prüfer: A combinatorial algorithm providing al-
     ternating approximations for a zero of an M-function,
     to appear in SIAM J. Num. Anal.
[10] M. Prüfer: Alternating approximations for solutions
     of nonlinear problems, to appear in ZAMM
[11] W.C. Rheinboldt: On M-functions and their applica-
     tions to nonlinear Gauss-Seidel iterations and to
     network flows, J. Math. Anal. Appl. 32 1970,
     pp. 274 - 307
[12] W.C. Rheinboldt and C.K. Mesztenyi: A combinatorial
     search process for M-functions, Beiträge zur nume-
     rischen Mathematik 4 1975, pp. 171 - 177
[13] J. Schröder: M-matrices and and generalizations
     using an operator theory approach, SIAM Review 20 2
     1978, pp. 213 - 244
[14] J. Vandergraft: Newtons method for convex operators
     in partially ordered spaces, SIAM J. Num. Anal. 4
     1967, pp. 406 - 432

AN EFFICIENT PROCEDURE FOR TRAVERSING LARGE PIECES

IN FIXED POINT ALGORITHMS*

R. Saigal

Northwestern University
Evanston, Illinois 60201

ABSTRACT

In this note we give an efficient procedure for traversing
larger pieces of linearity that result when the underlying functions have special structures including separability, bandedness,
linearity and piecewise linearity. It is shown here that the work
involved is a little more than that required to move through a
simplex. In particular, one may require $0(n \log_2 n)$ instead of $0(n)$
comparisons. The additions and multiplications are of the same
order. Since several simplexes lie in these larger pieces, substantial savings result.

1. INTRODUCTION

We consider the problem of computing an x in R^n such that
$f(x) = 0$, when $f: R^n \to R^n$ is a given continuous mapping. In particular, we consider the fixed point algorithms for solving this
problem. Two notable and early algorithms that can be used are
those of Eaves and Saigal[1] and Merrill[3].

Given a one-to-one affine mapping $r: R^n \to R^n$, in these algorithms a piecewise linear approximation G_ℓ to the homotopy
$G: R^n \times [0,1] \to R^n$ given by

$$G(x,t) = (1 - t)r(x) + tf(x) \tag{1.1}$$

---

*This research has been partially supported by the grant
MCS80-05154 from the National Science Foundation.

is implemented. Starting with the unique zero x_0 of r, a connected component of $G_\ell^{-1}(0)$ containing $(x_0,0)$ is traced, and success is achieved when a point $(x_1,1)$ in $G_\ell^{-1}(0)$ has been found. Then x_1 is an approximate solution to our problem.

Inherent to the process of tracing a component of $G_\ell^{-1}(0)$ is the triangulation of $R^n \times [0,1]$. Thus G_ℓ is linear in each $(n+1)$-dimensional simplex. Since r is affine, Todd[6] observed that the pieces of linearity of G_ℓ are in general unions of several simplexes, and considerable savings can result during the "path tracing" if care is taken to explicitly exploit this fact. Earlier, Kojima[2] had observed that if f is a separable mapping, i.e., there exist mappings $g_i: R \to R^n$, $i = 1,\ldots,n$, such that $f(x) = \sum_{i=1}^n g_i(x_i)$, the pieces of linearity of G_ℓ are again unions of many simplexes, and he presented a procedure to explicitly use this fact to reduce the computational effort. Todd[6] subsequently simplified this work and has pursued the study in several papers, including Todd[7]. Other instances when this occurs have also been identified by Todd[6].

The aim of this note is to give a simple procedure to trace the component of $G_\ell^{-1}(0)$ within a piece of linearity. This tracing is accomplished by identifying the expression for $G_\ell^{-1}(0) \cap \Sigma$, for a piece of linearity Σ. Σ, in general, is a union of several simplexes in the triangulation of $R^n \times [0,\Delta]$. The procedure presented here is based on the work of Saigal[4]. In contrast to our approach, both Kojima[2] and Todd[7] modify the system of equations which are solved and updated during the path tracing.

2. NOTATION, DEFINITIONS AND PRELIMINARY RESULTS

Throughout this note, we will assume that $R^n \times [0,1]$ is triangulated by K, Todd[5]. The simplexes of this triangulation are generated as follows: $\Delta > 0$ is a given real number, π a permutation of $\{1,2,\ldots,n+1\}$. Then v in R^{n+1} is called a vertex of K if and only if $v_i|\Delta$ is an integer, for each $i = 1,\ldots,n+1$. Now, each simplex σ in K can be represented by a pair (v,π) such that the vertices of σ are

$$v^1 = v$$
$$v^{i+1} = v^i + \Delta u_{\pi(i)}, \quad i = 1,\ldots,n+1, \tag{2.1}$$

where u_i is the ith unit vector in R^{n+1}. σ can also be expressed as the intersection of $(n+2)$-half spaces defined by: $x \in \sigma$ if and only if $y = -v + x$ satisfies

$$\Delta \geq y_{\pi(1)} \geq y_{\pi(2)} \geq \cdots \geq y_{\pi(n+1)} \geq 0. \tag{2.2}$$

Also, if σ and $\bar{\sigma}$ share a common n dimensional face τ with $\sigma = \tau \cup \{v\}$ and $\bar{\sigma} = \tau \cup \{\bar{v}\}$, then there are vertices u and w of σ such that

$$\bar{v} = u - v + w.$$

In particular, if $v = v^i$ for some i, then

$$\bar{v} = \begin{cases} \bar{v}^i = v^{i-1} - v^i + v^{i+1}, & 2 \le i \le n+1 \\ \bar{v}^1 = v^1 - v^{n+2} + v^{n+1}, & i = n+2 \\ \bar{v}^{n+2} = v^{n+2} - v^1 + v^2, & i = 1. \end{cases} \qquad (2.3)$$

Given a mapping $\ell: R^{n+1} \to R^n$, and a simplex σ in the triangulation K of R^{n+1}, we denote by A_σ the Jacobian of the linear approximation to ℓ on σ. It can be readily confirmed that if

$$C_\sigma = (\ell(v^1) - \ell(v^2), \ldots, \ell(v^{n+1}) - \ell(v^{n+2}))$$

when $\sigma = (v^1, \ldots, v^{n+2})$, then

$$A_\sigma V = C_\sigma \qquad (2.4)$$

where $V = (v^1 - v^2, \ldots, v^{n+1} - v^{n+2})$ is $-\Delta$ times the permutation matrix defined by π.

To implement homotopy (1.1) on a triangulation of $R^n \times [0, \Delta]$ by K, with a mesh size $\Delta > 0$, we define

$$\ell(x, t) = \begin{cases} f(x) & \text{if } t = \Delta \\ r(x) & \text{if } t = 0. \end{cases}$$

Then, it can be readily confirmed that a simplex $\sigma = (v^1, v^2, \ldots, v^{n+2})$ contains a zero of G_ℓ if and only if the system below has a solution:

$$\lambda_i \ge 0 \qquad\qquad i = 1, \ldots, n+2$$

$$\sum_{i=1}^{n+2} \lambda_i \, \ell(v^i) = 0 \qquad\qquad (2.5)$$

$$\sum_{i=1}^{n+2} \lambda_i = 1.$$

The system (2.5) naturally leads us to the matrix

$$D_\sigma = \begin{bmatrix} \ell(v^1) & \ell(v^2) & \cdots & \ell(v^{n+2}) \\ 1 & 1 & \cdots & 1 \end{bmatrix}$$

which we will call the label matrix associated with σ. We note that the following relationship exists between D_σ and C_σ:

$$D_\sigma M = \begin{bmatrix} C_\sigma & | & \ell(v^{n+2}) \\ -- & -|- & ----- \\ 0 & | & 1 \end{bmatrix} \tag{2.6}$$

where

$$M = \begin{bmatrix} 1 & & & & & \\ -1 & 1 & & & & \\ & -1 & 1 & & & \\ & & & \ddots & & \\ & & & & \ddots & \\ & & & & 1 & \\ & & & & -1 & 1 \end{bmatrix}$$

is an $(n+2) \times (n+2)$ matrix.

   In case σ is such that (2.5) has a solution, counting the number of variables and equations it can be readily established that its solution set is a 1 dimensional polyhedron, which starts and ends at the boundary of σ. Under the usual nondegeneracy assumption, this polyhedron intersects exactly two n-dimensional facets of σ. Using this special property, the fixed point algorithms trace a component of $G_\ell^{-1}(0)$ that contains $(x_0, 0)$ by generating a sequence of simplexes

$$\tau_0, \sigma_1, \tau_1, \sigma_2, \ldots, \tau_{k-1}, \sigma_k, \tau_k \tag{2.7}$$

such that τ_0 is the unique n-simplex containing $(x_0, 0)$, σ_i are $(n+1)$-simplexes for which (2.5) has a solution, and τ_{i-1} and τ_i are the two facets of σ_i which intersect $G_\ell^{-1}(0)$.

   In the usual implementation of the above path tracing, one usually stores the inverse of the matrix D_{τ_i}, where τ_i is the n-dimensional facet encountered in the algorithm. This inverse could be stored explicitly, or in several factored forms. When the mappings f and r have sparse Jacobians, it has been proposed, Saigal[4], that the inverse of C_{τ_i} be stored instead, since this matrix inherits the sparsity structure of the Jacobians. We will assume, for this work, that we have stored, instead, the inverse of D_{τ_i}. Given D_{τ_i} and σ_{i+1}, we now show how the usual implementations compute τ_{i+1} and σ_{i+2}. We assume that $\sigma_{i+1} = \tau_i \cup \{v\}$. The following is the well known procedure for continuing the path tracing:

Step 1: Calculate $\underline{\ell} = \left(\ell(v),1\right)^{T}$.

Step 2: Compute $\bar{\ell} = D_{\tau_i}^{-1}\,\underline{\ell}$.

Step 3: Determine the facet τ_{i+1} by a lexicographic minimum ratio test (as is usual in linear programming theory).

A geometric look at steps 1 through 3 is very revealing. Our aim is to trace the component $G_{\ell}^{-1}(0) \cap \sigma_{i+1}$, when it is known that $G_{\ell}^{-1}(0) \cap \tau_i \neq \emptyset$. Let $x = G_{\ell}^{-1}(0) \cap \tau_i$. Then, if $\tau_i = (u^1, u^2, \ldots, u^{n+1})$, using the solution λ_i to the resulting system (2.5), we can write

$$x = \sum_{i=1}^{n+1} \lambda_i\, u^i.$$

Let $\sigma = \sigma_{i+1}$. Since τ_i is a facet of σ, for some permutation matrix P, and $\hat{d} = P\bar{d}$, where $\bar{d} = (\bar{\ell}, -1)$,

$$D_{\sigma}\,\hat{d} = 0 \qquad\qquad (2.8)$$

or $\quad D_{\sigma} M \cdot M^{-1}\hat{d} = 0.$

Thus, using (2.6) and letting $\tilde{d}$ be the first $n+1$ components of $M^{-1}\hat{d}$, we note that

$$C_{\sigma}\,\tilde{d} = 0$$

and, thus, for $d = V\tilde{d}$, $A_{\sigma}d = 0$.

Now, let the $n+2$ bounding hyperplanes of σ be numbered by the order in which they appear in (2.2). Also, let the index of the hyperplane in which τ_i lies be s. Now, by possibly changing the sign of d, assure that $d_{\pi(s)} \leq \Delta$, if $s = 1$, or

$$d_{\pi(s-1)} \geq \begin{cases} d_{\pi(s)} & \text{if } s = 2, \ldots, n+1 \\ 0 & \text{if } s = n+2. \end{cases}$$

Then, it can be readily shown that $G_{\ell}^{-1}(0) \cap \sigma \subseteq \{x + td : t \geq 0\} = L$. Thus, the exiting face of σ can be easily found by computing $y = v - x$, and

$$t_j = \begin{cases} (y_{\pi(j)} - \Delta)/d_{\pi(j)}, & j = 1 \\ (y_{\pi(j)} - y_{\pi(j-1)})/(d_{\pi(j)} - d_{\pi(j-1)}), & j = 2, \ldots, n+1 \quad (2.8) \\ y_{\pi(j-1)}/d_{\pi(j-1)}, & j = n+2 \end{cases}$$

where t_j should be defined as $+\infty$ if the denominator is zero. The
line L then leaves σ through the face on the hyperplane with index
r, when $t_r = \min\{t_j : t_j > 0\}$.

3.   TRAVERSING LARGE PIECES

During the process of tracing the path (2.7), as developed in
the previous section, some properties of the function f result in
several simplexes of the path belonging to the same piece of linearity of G_ℓ. In this section we will investigate the possible
savings that may result. Essentially, for tracing the portion of
the path (2.7) which lies in the same piece of linearity of G_ℓ, the
steps 1-3 of the previous section specialize, and can be carried
out very efficiently.

These savings are achieved by observing that if the portion

$$\tau_{i_1-1}, \sigma_{i_1}, \ldots, \sigma_{i_2}, \tau_{i_2}$$

of the path (2.7) lies in the same piece of linearity Σ of G_ℓ, A_{σ_j},
$j = i_1, i_1+1, \ldots, i_2$ is the same matrix, and that $G_\ell^{-1}(0) \cap \Sigma =$
$\bigcup_{j=i_1}^{i_2} G_\ell^{-1}(0) \cap \sigma_j \subseteq L = \{x + td : t \geq 0\}$, where $x \in \tau_{i_1-1}$ and d is
such that $A_{\sigma_j} d = 0$ (as determined at the end of section 2). Our
idea now is the following. We will determine the exiting face τ_{i_2}
and a point $\tilde{x} \in \tau_{i_2} \cap L$ (without explicitly generating σ_{i_1},
$\sigma_{i_1+1}, \ldots, \sigma_{i_2}$) and thus continue into the piece adjacent to Σ by
determining σ_{i_2+1} and $D_{\tau_{i_2}}^{-1}$.

For the triangulation K, it can be readily confirmed that L
enters a new piece of linearity of G_ℓ if and only if it crosses
over one of the hyperplanes listed in Table 3.1.

Using the hyperplanes specified in Table 3.1, one can readily
determine the hyperplane in which τ_{i_2} lies. We now illustrate this
for the case when f is separable. Define $\hat{x} = v - x$, and for $i =$
$1, \ldots, n$

$$t_i = \begin{cases} \dfrac{\hat{x}_i - \hat{x}_{n+1}}{d_i - d_{n+1}} & \text{if } d_i - d_{n+1} \neq 0 \\[2mm] \infty & \text{otherwise} \end{cases}$$

and, for $i = 1, \ldots, n$

Table 3.1

Property of f	Hyperplanes bounding pieces of linearity of G_ℓ	Number of [1] Hyperplanes
Piecewise[2] Linear	$H_i = 0$, $i = 1,\ldots,T$ (Hyperplanes bounding the piece of linearity of f) $x_i = v_i$, $\quad i = 1,\ldots,n$ $x_i - x_{n+1} = v_i - v_{n+1}$, $\quad i = 1,\ldots,n$ $x_i = v_i + \Delta$, $\quad i = 1,\ldots,n$	$3n + T$
Separable and Linear	$x_i = v_i$, $\quad i = 1,\ldots,n$ $x_i = v_i + \Delta$, $\quad i = 1,\ldots,n$ $x_i - x_{n+1} = v_i - v_{n+1}$, $\quad i = 1,\ldots,n$	$3n$
Separable with band width k (separable when $x_i, x_{i+1}, \ldots, x_j$ for $\|i-j\| \le k$ are constant). $n > 4k$.	$x_i = v_i$, $\quad i = 1,\ldots,n$ $x_i = v_i + \Delta$, $\quad i = 1,\ldots,n$ $x_i - x_j = v_i - v_j$, for all $\|i-j\| \le 2k$, $\quad j = 2,\ldots,n$ $\quad$ for all $i < j = n+1$	$3n + 4nk$ $- 2k - 4k^2$
Partial Separability- (separable in $x_{k+1},\ldots,x_n$ when $x_1 \cdots x_k$ are fixed)	$x_i = v_i$, $\quad i = 1,\ldots,n$ $x_i = v_i + \Delta$, $\quad i = 1,\ldots,n$ $x_i - x_j = v_i - v_j$, for all $i < j \le k$ $\quad$ for all $i \le k$, $\quad j \ge k+1$ $\quad$ for all $i < j = n+1$	$3n + k(n-k)$ $+ \dfrac{k(k-1)}{2}$

These inequalities have been written for the triangulation K, and $\sigma_{i_1} = (v, \pi)$.

$$
t_{i+n} = \begin{cases} \dfrac{\hat{x}_i + \Delta}{d_i} & \text{if } d_i < 0 \\ \infty & \text{if } d_i = 0 \\ \dfrac{\hat{x}_i}{d_i} & \text{if } d_i > 0. \end{cases}
$$

Now, define

$$
t_m = \min_{1 \le i \le 2n} \{t_i | t_i > 0\}.
$$

[1] All these hyperplanes do not necessarily bound a particular piece Σ of linearity of G_ℓ.

[2] We assume that the boundary of a piece of linearity of f does not lie interior to any simplex of K.

Since the piece Σ is bounded, t_m exists. Also, we observe that τ_{i_2} lies in the hyperplane

$$x_i - x_{n+1} = v_i - v_{n+1} \qquad \text{if } i = m \leq n$$

or $\qquad\qquad x_i = \text{constant} \qquad \text{if } i = m - n > 0.$

The justification for the above observation lies in the fact that t_i is the value for which $x + t_i d$ lies in the hyperplane specified by i. Thus τ_{i_2} is the facet of the simplex containing the point $\tilde{x} = x + t_m d$. For the triangulation K, we can find all the simplexes containing $\tilde{x}$ as follows:

Let $\hat{y} = -\hat{x} + t_m d$. Now, define a permutation $\bar{\pi}$ such that

$$\Delta \geq \hat{y}_{\bar{\pi}(1)} \geq \hat{y}_{\bar{\pi}(2)} \geq \cdots \geq \hat{y}_{\bar{\pi}(n+1)} \geq 0. \tag{3.1}$$

Then the point $\tilde{x}$ lies in the simplex $\bar{\sigma} = (v, \bar{\pi})$. In case $\tilde{x}$ lies interior to τ_{i_2}, then exactly one inequality in (3.1) will be an equality, i.e., for some r

$$\hat{y}_{\bar{\pi}(r)} = \hat{y}_{\bar{\pi}(r+1)}$$

then, if $d_{\bar{\pi}(r)} < d_{\bar{\pi}(r+1)}$, $\sigma_{i_2} = (v, \bar{\pi})$ and if $d_{\bar{\pi}(r)} > d_{\bar{\pi}(r+1)}$, $\sigma_{i_2+1} = (v, \bar{\pi})$.

In the contrary case, one would need a more expensive perturbation scheme to find the piece adjacent to Σ.

We now show how to generate the matrix of the inverse of labels, $D^{-1}_{\tau_{i_2}}$, associated with the facet τ_{i_2}, which is needed during the steps $1 - 3$ to determine the new "direction" d into the piece of linearity adjacent to Σ.

We also assume that we have the label matrix $D^1 = D_{\sigma_{i_1}}$ and the partitioned form (2.6). We note that in triangulation K all simplexes belonging to the same hypercube share the vertices $v^1 = v$ and v^{n+2}. Thus

$$D^2 M = \left[\begin{array}{c|c} C^2 & \ell(v^{n+2}) \\ \hline 0 & 1 \end{array} \right] \tag{3.2}$$

where $C^2 = C_{\sigma_{i_2}}$, $D^2 = D_{\sigma_{i_2}}$. Also, using (2.4), since $A_{\sigma_{i_1}} = A_{\sigma_{i_2}}$, we can write

$$C^2 = C^1 V_1^{-1} V_2 \tag{3.3}$$

where $V_i = V_{\sigma_{i_i}}$, $i = 1, 2$.

Thus, if $N_i = \begin{bmatrix} -\frac{1}{\Delta} V_i & \vdots & 0 \\ - - - - & - \vdots - - \\ 0 & \vdots & 1 \end{bmatrix}$, $i = 1, 2$ (for the triangulation

K, N_1 and N_2 are permutation matrices), using (3.2) and (3.3) we note that

$$D2 = D1 \, M N_1^{-1} \, N_2 M^{-1}.$$

Given $D_{\tau_{i_1-1}}^{-1}$, we can readily define a matrix B^1, by inserting a row of all zeros in $D_{\tau_{i_1}}^{-1}$ at the position of the vertex of σ_{i_1} not in τ_{i_1-1}.

We also note that

$$D^1 \, B^1 = I. \tag{3.4}$$

From (2.8), we also have

$$D^1 \, \hat{d}^1 = 0. \tag{3.5}$$

Thus, $B^2 = M N_2^{-1} \, N_1 M^{-1} \, B^1$

$\hat{d}^2 = M N_2^{-1} \, N_1 M^{-1} \, \hat{d}_1$

satisfy (3.4) and (3.5) respectively for D^2. Thus $D_{\tau_{i_2}}^{-1}$ is obtained by eliminating the all zero row r from the matrix

$$B^2 - \frac{1}{\hat{d}_r^2} \, \hat{d}^2 \, B_r^2$$

where the vertex in σ_{i_2} not in τ_{i_2} is v^r.

For a separable function, the total work involved in pivoting through Σ can be summarized as

Table 3.2

Operation	Comparisons	Multiplications	Additions
Determining L	–	$(n+1)^2$	$(n+1) + (n+1)^2$
Pivoting through Σ	$3n+1$ $+ (n+1)\log_2(n+1)$	$3n+2$	$5n+2$
Generating $D_{\tau_{i_2}}^{-1}$	–	$(n+1)^2 + (n+1)$	$3(n+1)^2$

We now compare Table 3.2 to the work involved in generating one simplex on the path (2.7) when the function f has no special structure. In this case, one can pivot through a simplex by a minimum ratio test involving $n+1$ comparisons and $n+1$ divisions. The other operations are of the same order of magnitude. Thus the additional effort involved is an order of magnitude more comparisons ($O(n \log_2 n)$ instead of $O(n)$). Thus, this scheme is very effective in cases where a large number of simplexes make up a single piece of linearity Σ of G_ℓ.

REFERENCES

1. B. C. Eaves and R. Saigal, Homotopies for computation of fixed points on unbounded regions, Mathematical Programming, 3(1972), 225-237.

2. M. Kojima, On the homotopic approach to systems of equations with separable mappings, Mathematical Programming Study, 7 (1978), 170-184.

3. O. H. Merrill, Applications and Extensions of an algorithm that computes fixed points of certain upper semi-continuous point to set mappings, Ph.D. Dissertation, Department of Industrial Engineering, University of Michigan, Ann Arbor, Michigan, 1972.

4. R. Saigal, A Homotopy for Solving Large, Sparse and Structured Fixed Point Problems, Department of Industrial Engineering and Management Sciences, Northwestern University, Evanston, Illinois, Jan. 1981.

5. M. J. Todd, "The Computation of Fixed Points," Springer-Verlag, New York, 1976.

6. M. J. Todd, Exploiting Structure in Piecewise Linear Homotopy Algorithms for Solving Equations, Mathematical Programming, 18 (1980), 233-247.

7. M. J. Todd, Traversing Large Pieces of Linearity in Algorithms that Solve Equations by Following Piecewise Linear Paths, Mathematics of Operations Research, 5 (1980), 242-257.

THE APPLICATION OF FIXED POINT

METHODS TO ECONOMICS*

John B. Shoven

Department of Economics
Stanford University
Stanford, California 94305

The purpose of this paper is to briefly review the applications of fixed point and path following methods to problems in economics, and to present one such application in some detail. The problem for which economists have adopted these methods is the computation of a set of goods prices which represent an economy's equilibrium. By equilibrium prices, we mean that total market demands match total market supplies for every good, when both consumers and producers maximize their objective functions subject to these prices. Methods of the type described in the other papers in this conference volume are proving to be extremely useful for solving this difficult economic problem. Perhaps economists have accepted these techniques more rapidly than others because of the pathbreaking work in the computation of fixed points by an economist, Herbert Scarf. Scarf's original articles [1967a, 1967b] and landmark book [1973] deal with the computation of economic equilibria.

Scarf's publications, as well as other contributions to the literature on computational algorithms, usually include simple illustrative examples. These examples are neither meant to capture the actual structure of any real economies nor to evaluate the impact of new economic policies. However, in recent years the economics profession has adapted these computational devices for real empirical research, and it is the nature of these applications which this paper attempts to describe.

The original Scarf algorithm has a severe drawback in applications where a precise equilibrium is desired. It is extremely

---

*The author is indebted to Charles Ballard for generous research assistantship on this paper and on the work which it summarizes.

costly to compute a close approximation to an equilibrium, due to
the lack of a re-start capability. Scarf's algorithm uses a fixed
simplex grid, which cannot be refined, and it begins in a corner of
the simplex. Therefore, it spends considerable time simply getting
to the vicinity of an equilibrium. This problem was essentially
solved by the re-start algorithms of Merrill [1972] and Kuhn [1968].

At this time, the available solution algorithms seem adequate
to inexpensively solve most of the economic models which have been
constructed. The limiting factor has become the need to collect
appropriate data sets and deal with a myriad of modeling issues. An
unusual combination of skills is required to do research in this area.
Applied general equilibrium modeling requires a substantial knowledge
of mathematical economics and computer programming, and a willingness
to deal with large and often inconsistent data sources. The available
computer programs are far from being "off-the-shelf" items at this
point. No single course in economics, operations research, or statistics will equip a student to conduct research in this area.

Despite the relatively high "barriers of entry" to this field,
a significant number of studies have been completed. These include
a large number of models of specific countries, such as Andrew
Feltenstein's [1980] work on Argentina, the large model of Australia
developed by Peter Dixon et al. [1982], and Shujiro Urata's [1978]
dissertation on the Japanese economy. Other examples include my work
with Don Fullerton, John Whalley, Lawrence Goulder, Charles Ballard
and others on the United States economy (Fullerton et al., [1981a],
[1981b], [1982], and Goulder et al., [1981]), John Piggott and John
Whalley's book [1981] on the British economy, and, finally, the work
of Jaime Serra-Puche and Timothy Kehoe on Mexico (Serra [1979] and
Kehoe and Serra [1981]). Most of these models are used to evaluate
alternative domestic tax policies. In each case the objective is to
compare the economy's equilibrium under a new policy with the situation in its absence. The models determine a new set of equilibrium
prices in the counterfactual case of implementing the new policy.
Corresponding to these prices is a complete set of consumer and
producer activities. In many cases, the investigators attempt to
measure the change in the level of economic welfare for particular
groups in the economy as a result of the new policy, as well as for
the economy as a whole. For the most part, the single country tax
models are static and the results are to be considered as medium
to long run forecasts. The U.S. model of Fullerton et al. permits
dynamic evaluations of tax policies.

Other researchers have used applied general equilibrium models
in other areas. Edward Hudson and Dale Jorgenson [1974] have done
extensive work on energy models, as have Antonio Borges and Lawrence
Goulder [1981]. Jorgenson [1981] has also led the way on econometri-
cally estimating general equilibrium models, and is among the handful
of authors who are developing dynamic versions of the approach.

James MacKinnon [1974] has used these techniques for urban economics
issues and also has done work in the analysis of alternative tax
policies.   Larry Kimbell and Glenn Harrison [1981] have worked on
models of intergovernmental relations and have done some work on
developing algorithms which are particularly efficient for their type
of problem.   Several people have worked in the international trade
area (including John Whalley [1980] and Victor Ginsburgh and Jean
Waelbroeck [1975]).   Finally, several authors have been building
general equilibrium models of developing countries.   These include
Sherman Robinson and Laura D'Andrea Tyson [1981] and Kemal Dervis,
Jaime deMelo, and Robinson [1981].

     The structure of the various models depends upon the problems
addressed.   Understandably, Hudson and Jorgenson and Borges and
Goulder model the energy sectors in greater detail than do other
researchers.   Piggott and Whalley deal carefully with housing subsi-
dies, which are important in the U.K.   The U.S. model is designed to
analyze tax policy, so it has a greater variety of taxes than many of
the·other models.

     Clearly, the above models differ in detail, but they also bear
familial similarities.   First, they are all general rather than
partial equilibrium models.   The interactions of all markets are
incorporated.   Second, they assume competition, profit maximization,
and complete information on the part of all participants in the
economy.   Third, with the sole exception of Kehoe-Serra, they assume
full employment and complete mobility of factors between production
sectors.   Many of the models make a sharp distinction between factors
of production (labor and capital) and outputs.   In many cases the
problems can be computed in terms of equilibrium factor prices, which
means that the basic dimensionality is low.

     Because I am most familiar with the U.S. tax model of Fullerton,
Shoven, Whalley, and Ballard, I will survey its structure.   It is
quite representative of this type of applied general equilibrium model.
It is designed for long-run evaluation of alternative United States
federal tax policies.   Its structure is shown in Figure 1.   There are
19 production sectors in the economy, each of which uses capital (K)
and labor (L) and the outputs of the other 19 industries in the production of output.[1,2]   The firm is able to substitute capital and

---

[1]The industries are Agriculture, Forestry, and Fisheries;
Mining; Crude Petroleum and Gas; Contract Construction; Food and
Tobacco; Textiles, Apparel, and Leather Products; Paper and Printing;
Petroleum Refining; Chemicals and Rubber; Lumber, Furniture, and
Stone; Metals, Machinery, and Miscellaneous Manufacturing; Transpor-
tation Equipment; Motor Vehicles; Transportation, Communications and
Utilities; Trade; Finance and Insurance; Real Estate; Services; and
Government Enterprises.

[2]Some applied general equilibrium models are substantially

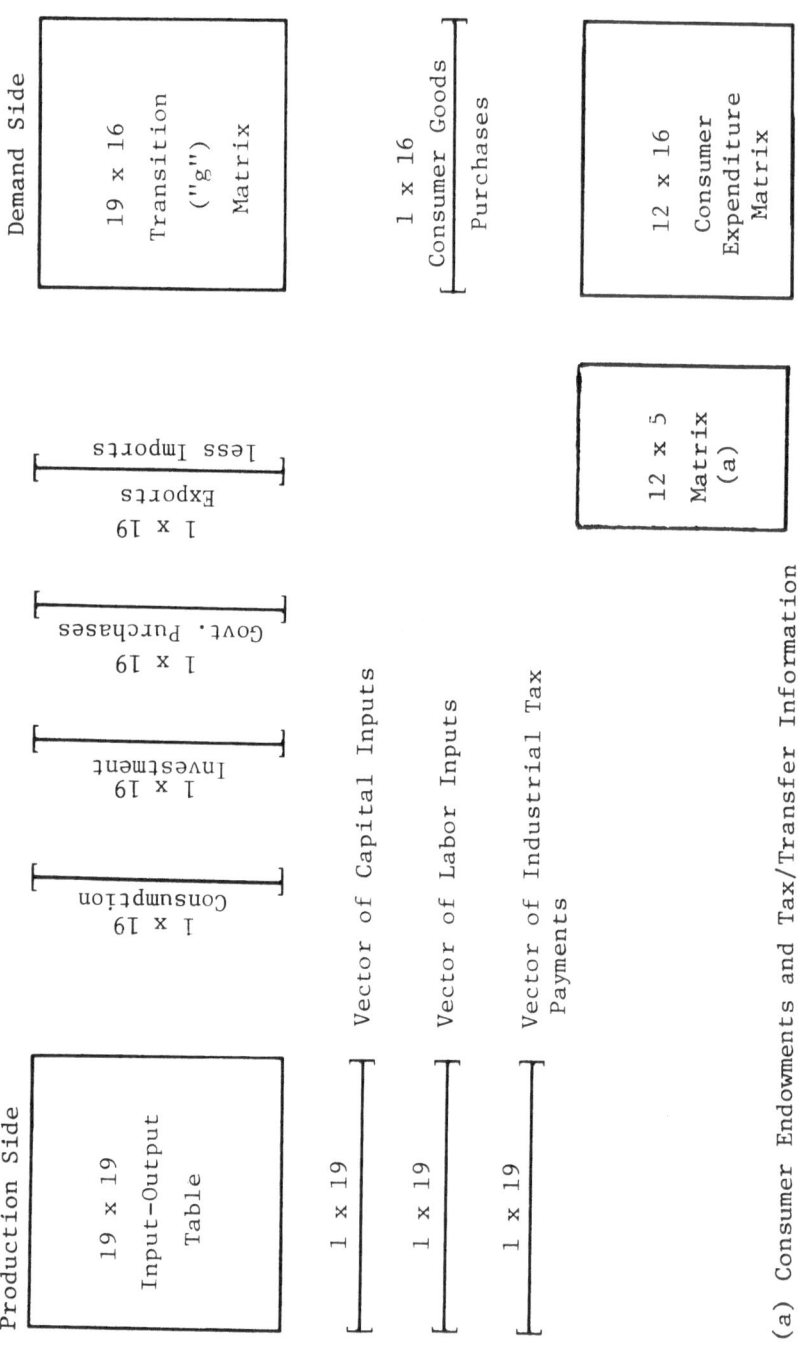

Figure 1.   Size and Structure of U.S. Model.

and labor in production, but must use the outputs of the other sectors in fixed proportions. That is, each sector j has a production function of the form

$$Q_j = \min\left[\frac{VA_j(K_j, L_j)}{a_{0j}}, \frac{X_{ij}}{a_{ij}}, \ldots, \frac{X_{nj}}{a_{nj}}\right]$$

where Q_j is the gross output of jth industry, VA_j is the value-added in the sector, and X_{ij} is the input of output i into the production of output j. The numbers $a_{0j}, \ldots, a_{nj}$ are the fixed input-output coefficients, n is the number of production sectors (19 in this case), and thus the input-output coefficients form the 19 x 19 matrix shown in Figure 1. The value-added function is usually specified in constant elasticity of substitution (CES) form, and looks like

$$VA_j = \gamma_j\left[\alpha_j L_j^{-\rho_j} + (1-\alpha_j)K_j^{-\rho_j}\right]^{-1/\rho_j}.$$

The gross production output of each of the 19 sectors is used for government purchases, investment or capital formation, net exports, intermediate inputs, and the creation of consumer goods. The 19th production sector is somewhat different from the others in that it consists of government enterprises (Tennessee Valley Authority, U.S. Postal Service, etc.) and is subsidized rather than taxed by the government.

Consumer goods are classified differently than producer goods. This simply reflects the way in which the data are collected. Production data include classifications such as chemicals and rubber, whereas expenditure data list such items as household appliances and auto repair. In the U.S. model there are 16 consumer goods[3] which are produced from the 19 industrial sectors. The technology of conversion between industrial output and consumer goods is assumed to be characterized by fixed coefficients. The data for the conversion are gathered in the 19 x 16 "g" matrix of Figure 1. The last, or 16th,

---

more disaggregated. For example, the model of Dixon et al. includes more than 100 sectors, while the Hudson-Jorgenson model has 36 sectors.

[3]The consumer goods are Food; Alcoholic Beverages; Tobacco; Utilities; Housing; Furnishings; Appliances; Clothing and Jewelry; Transportation; Motor Vehicles, Tires, and Auto Repair; Services; Financial Services; Reading, Recreation, and Miscellaneous; Nondurable-nonfood Household Items; Gasoline and Other Fuels; and Savings.

consumer good is "saving." It is somewhat different from the others.
The consumers can purchase this commodity, which is actually a fixed
coefficient portfolio of investment goods, and thereby augment their
capital endowments and increase their provision for purchases in the
future.

The U.S. tax model classifies consumers into 12 income groups
on the basis of 1973 income. Each group has endowments of capital
and labor. Each group makes choices between work and leisure and
between savings and consumption. Each group faces income taxes and
is eligible for transfer programs such as welfare or Social Security,
and each must decide how to allocate its consumption expenditures
between the commodities in the model. The consumer groups are
modeled as constrained utility maximizers. The constraint is that
the consumers cannot spend more than the market value of their endowments, plus transfer payments, less taxes. Each consumer has a
utility function. The consumers may have different parameters in
their utility functions, but all have the same nested form for the
function. The form is

$$U = U\left[H\left(\prod_{i=1}^{m} C_i^{\lambda_i}, L\right), C_f\right],$$

where C_i is the consumption of the ith consumer good (not including

savings), C_f is the future consumption which the consumer expects to

receive due to savings, and L is leisure. m is the number of consumer
goods other than savings and is 15 in the present U.S. model. The H
function, which could be termed the "happiness" function, is of the
CES type, as is the U function. In parameterizing these functions in
the U.S. model, two key elasticities are set. The elasticity of
savings with respect to the real after-tax rate of return is set at
0.4 (meaning that a doubling of the real rate will increase savings
by 40 percent) and the labor supply elasticity with respect to the
wage rate is pegged at 0.15. The economics profession has not
actually reached precise agreement on these key statistics, but these
seem to represent reasonable values, which are roughly in the center
of those estimated in recent studies.

The model is designed for tax policy evaluation, and hence
includes a complete array of taxes. These include the corporate
income tax, the corporation franchise tax, property taxes, sales
taxes, excise taxes, Social Security taxes, and the personal income
tax. The government collects revenue from these sources and uses
the resources to purchase goods and services (G), to finance transfer and welfare programs, and to subsidize government enterprises.
As currently structured, the government always runs a balanced
budget in this model. Future developments may permit deficits,

which would be feasible if a government bond market were added to the model.

Figure 1 not only shows the size and structure of the model, but also gives an accurate picture of the data requirements. A problem which all researchers in this area encounter is that data are not readily available in this format. The data must be gathered from many different sources and will, in general, be inconsistent. For example, the total payments to labor (gathered from one data source) will not be equal to the receipts of labor income (from another source). This problem is widespread in the data collection. One must determine which numbers are most plausible and then adjust others to be consistent. Some of the required information is partic- ularly difficult to obtain. For instance, we need to know capital utilization by industry. Almost all data sources report capital ownership by industry. It causes difficulty if, for instance, the insurance industry owns airplanes which they then lease to airlines. The data required for this model would deal with the industry in which the capital is utilized (airlines), rather than the industry which owns the capital, but this unscrambling is a tedious business. All in all, collecting the data for one year for a U.S. model of this size involves between six months and one year of full time work.

The next step in this general equilibrium simulation approach is to parameterize the production and utility functions in the model. Here we use a technique which could be termed "replication." We have a complete and (by this stage) consistent data set. We assume that the data represent an equilibrium.[4] That is, for current tax poli- cies, we know the "answer" the model is supposed to generate. Consequently, we can attempt to solve the model "backwards." If we know that the model generated the existing data, and if we assume that the economy was in equilibrium, this helps us to determine what the parameters of the model must have been. In general, the data are insufficient to identify the model uniquely, but forcing the model to reproduce the data does reduce the parameterization problem consider- ably. If the two utility function elasticities mentioned, savings and labor supply, are set, then the requirement of data replication is sufficient to determine the parameters of the utility functions uniquely. Similarly, if one sets the parameter which specifies the ease with which each industry can substitute capital for labor in its value-added function, then the data allow one to determine the remain- ing parameters of the production and value-added functions.

---

[4]This assumption is necessary to specify the model in the non- econometric manner described here. While 1973 was an inflationary year, the real economy in the U.S. was probably close to full employ- ment and had yet to face the oil shock that accompanied the formation of OPEC.

The U.S. tax model is one of the few that have begun to deal with dynamic adjustments to new policies. It calculates a sequence of equilibria, which are connected to each other by exogenous labor force growth and endogenous capital growth. The growth of capital is determined by savings in this essentially classical model. We assume that the economy is growing in a balanced manner before the new policy is introduced (i.e., capital and labor grow at the same rate). We calculate the transition to the new growth path, induced by the tax policy change.

Because of this dynamic structure, one must model the expectations on which consumers base their decisions. In the U.S. tax model, these expectations are myopic. Consumers assume that today's relative prices will exist in the future. They assume that their savings will earn the same rate of return that today's capital stock is earning. The myopic expectations assumption contrasts with the assumption of perfect foresight, which is adopted by Auerbach, Kotlikoff, and Skinner [1981]. Ballard and Goulder [1982] have done some work on comparing these two extreme expectational structures, as well as intermediate degrees of foresight, within the framework of the U.S. tax model. They find that perfect foresight does not produce results which are very different from those which stem from myopic expectations. In addition, it appears that a small degree of foresight (e.g., 10 years) is sufficient to yield most of the changes which would occur if we moved all the way to perfect foresight.

The effects of a tax policy revision are measured by comparing two sequences of equilibria. The base sequence begins with the 1973 data, and we assume that it simply grows in a balanced manner, with relative prices remaining constant. In the revised sequence (after a tax policy change), relative prices will adjust for a number of equilibria, but will eventually stabilize when a new balanced growth situation has been achieved. Much of our work has dealt with changes in the manner in which capital income is taxed. Investors always reallocate their capital portfolio so that they receive the same after-tax return from capital from every industry. Because of this, differential taxes across industries will cause capital to be misallocated, in the sense that it will be socially more productive in some industries than in others. Economic welfare could be enhanced if capital were moved to its most productive locale. This requires that the last unit of capital be equally productive in all locations.

When the model compares the existing tax system with an alternative, it determines a single scalar at each point in time which allows one to scale the new tax system to match the government revenue generated by the existing tax code. This "equal yield" feature permits a more realistic and unbiased comparison of alternative tax programs.

Our model is made more complicated by the presence of government revenue. In models with no taxes, prices convey all the information that consumers and producers need in order for them to make decisions about consumption, savings, leisure, and production. With taxes, the situation is more complicated, because these decisions are made at the same time that government revenue is collected. To see the issue most clearly, assume that the government returns all tax proceeds to consumers as transfer payments. Now, a consumer cannot know his income (and hence determine his consumption) until the government's revenue is announced. However, the government's revenue depends on tax collections, which are determined by economic activities, including consumption. The solution is to work computationally with one extra simplex dimension, and to announce government revenue on which decisions are based. In equilibrium, the announced government revenue will indeed be collected.

The development of this tax model was financed by the U.S. Treasury Department, which now uses the model to evaluate the long-run implications of possible revisions in the tax code. It has already been used for a number of studies. These include the unification or integration of the corporate and personal income tax codes, the introduction of additional incentives for saving (including the idea of taxing consumption rather than income), changing depreciation policies (an idea which became reality in 1981), and an examination of foreign tax credits. In this paper, I can only report a small sampling of the results obtained and cannot properly qualify them. Nonetheless, a summary of some findings is given in Table 1.

What Table 1 shows is that both integrating the corporate and personal income tax systems and moving to the taxation of consumption rather than income offer substantial gains in economic efficiency. The two policies are evaluated separately and in conjunction. The numbers reported are the present value of the infinite stream of efficiency gains (or losses) which result from the new policy, where the real discount rate is four percent. With the policies shown in Table 1, government revenue would fall if no adjustments are made elsewhere in the tax system. The table shows two alternative tax sources to replace the foregone revenue. The first is a non-distortionary lump sum ("head") tax while the second raises consumers' marginal tax rates by a flat number of percentage points sufficient to restore the government's revenue. The cases I would choose as central are those with the saving elasticity set at 0.4, the labor elasticity at 0.15 and the replacement revenues raised with higher marginal rates. For this set of assumptions, integrating the two tax systems involves a dynamic efficiency gain of $449 billion in 1973 dollars (about .9 percent of the present value of future income), adopting a consumption tax results in a $636 billion gain (1.3 percent) and both policies together give a boost in efficiency of $1,135 billion. The range of values for the elasticity parameters shown span the reasonable estimates. While the results depend on

Table 1.   Increase in Economic Efficiency
(measured in $ billion 1973)

1. Integrating Corporate and Personal Income Tax Systems
    (effectively eliminates the corporate tax)

Replacement Tax	Labor Supply Elasticity	Saving Elasticity w.r.t. Real Rate of Return		
		0.0	0.4	0.8
Lump Sum	0.0	$455	$693	$899
	0.15	$466	$732	$960
Addition to Consumer's Marginal Tax Rates	0.0	$261	$510	$727
	0.15	$160	$449	$707

2. Adoption of a Consumption Tax

Replacement Tax	Labor Supply Elasticity	Saving Elasticity w.r.t. Real Rate of Return		
		0.0	0.4	0.8
Lump Sum	0.0	$463	$618	$760
	0.15	$510	$686	$841
Addition to Consumer's Marginal Tax Rates	0.0	$416	$586	$742
	0.15	$438	$636	$817

3. Both Consumption Tax and Corporate Tax Integration

Replacement Tax	Labor Supply Elasticity	Saving Elasticity w.r.t. Real Rate of Return		
		0.0	0.4	0.8
Lump Sum	0.0	$929	$1311	$1659
	0.15	$993	$1430	$1818
Addition to Consumer's Marginal Tax Rates	0.0	$720	$1143	$1543
	0.15	$633	$1135	$1601

these values, the policies involve large efficiency gains for all parameter combinations. Briefly, the reason these tax reforms are beneficial to economic efficiency is that they remove the disincentives to save and to invest in the corporate sector which exist in the present U.S. tax system.

The computational costs of this model are modest. We have been able to take advantage of the fixed coefficient nature of the production functions to reduce the dimensionality of the computation, but we have not optimized the computer code in a systematic way. The cost of computing a sequence of ten equilibria spaced five years apart, and evaluating the terminal stock, comparing the sequence to the base policy path, etc., is about one dollar on the Stanford IBM 3033. The equilibrium is precise to at least six significant figures, which is far beyond that necessary for policy analysis purposes. These results are obtained using Merrill's algorithm.

The tax model can be improved in several dimensions. Additional disaggregation among consumers would be desirable, as would modeling their life cycle decisions. This work would permit one to evaluate the Social Security program more accurately. The production functions would probably be improved if they incorporated more substitution possibilities and dealt with technical change. The incentive characteristics of government transfer programs could likewise be modeled more realistically. I believe we currently have a model which is useful for policy purposes, but one which could be enhanced along several of these lines.

This type of model should prove useful in future evaluations of a wide array of economic problems. Some natural applications are an assessment of the minimum wage, the impact of unions on the allocation of labor and capital, and the value of government regulation programs. This approach is also the theoretically correct way to do cost-benefit studies. The usual practice is to ignore price changes, but this is clearly inappropriate for large programs such as fusion research, or the development of solar energy stations in outer space. The computational general equilibrium approach is also applicable to an assessment of international economic unions (such as the EEC or a possible North American trade union) and evaluation of alternative policies dealing with the energy "crisis." It is my opinion that, as the barriers to entry to this field are gradually lowered, as econometric techniques are developed for model estimation, and as economists get more sophisticated in terms of both computers and mathematics, the computational general equilibrium techniques will be a leading tool for empirical economic evaluation.

References

Auerbach, Alan J., Kotlikoff, Laurence J., and Skinner, Jonathan, 1981, The efficiency gains from dynamic tax reform. NBER Working Paper No. 819.

Ballard, Charles L., and Goulder, Lawrence H., 1981, Expectations in numerical general equilibrium models. Mimeo, Stanford University.

Borges, Antonio M., and Goulder, Lawrence H., 1981, Decomposing the impact of higher energy prices on long-term growth. Presented at the NBER Conference on Applied General Equilibrium Modeling, August 24-28, 1981. Forthcoming as an NBER Conference volume.

Dervis, Kemal, deMelo, James, and Robinson, Sherman, 1981, "General Equilibrium Models for Development Policy," Cambridge, Cambridge University Press.

Dixon, Peter B., Parmenter, B. R., Sutton, J., and Vincent, D. P., 1982, "ORANI: A Multisectoral Model of the Australian Economy," Amsterdam, North-Holland Publishing Co.

Feltenstein, Andrew, 1980, A general equilibrium approach to the analysis of trade restrictions, with an application to Argentina, International Monetary Fund Staff Papers, 27, 747-84.

Fullerton, Don, Shoven, John B., and Whalley, John, 1981a, Dynamic general equilibrium impacts of replacing the U.S. income tax with a progressive consumption tax. (An earlier version appeared as NBER conference paper No. 55.)

Fullerton, Don, King, A. Thomas, Shoven, John B., and Whalley, John, 1981b, Corporate tax integration in the United States: A general equilibrium model, American Economic Review 71, 677-91.

Ginsburgh, Victor, and Waelbroeck, Jean, 1975, A general equilibrium model of world trade, part I: Full format computation of economic equilibria, Cowles Foundation Discussion Paper 412, Yale University.

Goulder, Lawrence H., Shoven, John B., and Whalley, John, 1981, Domestic tax policy and the foreign sector: The importance of alternative foreign sector formulations to results from a general equilibrium tax analysis model. Presented at the NBER Tax Simulation Conference, January 26-27, 1981, Palm Beach, Fla. Forthcoming as an NBER Conference volume.

Hudson, Edward A., and Jorgenson, Dale W., 1974, U.S. energy policy and economic growth, Bell Journal of Economics and Management Science, 87, 523-43.

Jorgenson, Dale W., 1981, Econometric methods for applied general equilibrium modeling. Presented at the NBER Conference on Applied General Equilibrium Modeling, August 24-28, 1981. Forthcoming as an NBER Conference volume.

Kehoe, Timothy J., and Serra-Puche, Jaime, 1981, The impact of the 1980 fiscal reform on unemployment in Mexico. Presented at the NBER Conference on Applied General Equilibrium Modeling, August 24-28, 1981. Forthcoming as an NBER Conference volume.

Kimbell, Larry, and Harrison, Glenn, 1981, General equilibrium analysis of regional fiscal incidence. Presented at the NBER Conference on Applied General Equilibrium Modeling, August 24-28, 1981. Forthcoming as an NBER Conference volume.

Kuhn, Harold W., 1968, Simplical approximation of fixed points, Proceedings of the National Academy of Sciences, USA, 61, 1238-42.

MacKinnon, James, 1974, Urban general equilibrium models and simplical search algorithms, Journal of Urban Economics, 1, 161-83.

Merrill, O. H., 1972, Applications and extensions of an algorithm that computes fixed points of certain upper semi-continuous point-to-set mappings. Unpublished Ph.D. dissertation, University of Michigan.

Piggott, John, and Whalley, John, 1981, Economic effects of the U.K. tax subsidy policies: A general equilibrium approach, mimeo. To be published by Macmillan Publishing Co.

Robinson, Sherman, and Tyson, Laura D'Andrea, 1981, Modeling structural adjustment: Micro and macro elements in a general equilibrium framework. Presented at the NBER Conference on Applied General Equilibrium Modeling, August 24-28, 1981. Forthcoming as an NBER Conference volume.

Scarf, Herbert, 1967a, The approximation of fixed points of a continuous mapping, SIAM Journal of Applied Mathematics 15, 1328-43.

_____, 1967b, On the computation of equilibrium prices, in: "Ten Essays in Honor of Irving Fisher," Fellner, et al., eds., New York, John Wiley and Sons.

Scarf, Herbert with the collaboration of Terje Hansen, 1973, "The Computation of Economic Equilibria," New Haven, Yale University Press.

Serra-Puche, Jaime, 1979, A computational general equilibrium model for the Mexican economy: An analysis of fiscal policies. Unpublished Ph.D. dissertation, Yale University.

Urata, Shujiro, 1978, Effects of protection on resource allocation and economic welfare: The Japanese case. Unpublished Ph.D. dissertation, Stanford University.

Whalley, John, 1980, Discriminating features of domestic factor tax systems in a goods mobile-factors immobile trade model: An empirical general equilibrium approach, Journal of Political Economy, 88, 1177-1202.

ON A THEORY OF COST FOR EQUATION SOLVING

Mike Shub

Mathematics Department
Queens College
New York, N.Y.

Steven Smale

Mathematics Department
UC Berkeley
Berkeley, CA

We study algorithms for finding a zero of a single complex
polynomial, from the point of view of computational complexity.
Thus the problem is to give comparative assessment of the cost of
various methods for equation solving. The case of a single polynomial is focused on as the prototype of a non-linear system of
equations.

A class of algorithms which we call incremental algorithms is
studied systematically, and a subclass is distinguished for its
speed. Criteria are developed to minimize the cost relative to the
degree of the polynomial and probability of success.

The incremental algorithms are iterative ones and include the
classic and fast methods that have proved successful in practice.
A measure of efficiency is described and incremental algorithms of
efficiency k are characterized. Among these are a special series,
incremental k^{th} Euler algorithms, $k = 1, 2, 3, \ldots, \infty$, which
seem simplest to administer. For a method based on these algorithms we show that the number of steps required is linear in the
degree of the polynomial, and the number of multiplications quadratic. In view of the robustness, and that one has quadratic
order or higher near the solution, these estimates are quite
strong. The speed is comparable to Schur-Cohn methods, but for
Schur-Cohn, round-off errors are serious and convergence at a
solution is only linear.

Background for this paper is S. Smale, "The Fundamental
Theorem of Algebra and Complexity Theory," Bull. Amer. Math. Soc.,
1981, pp. 1-36. As there, we use heavily the theory of schlicht

functions. A full detailed account is in the process of being written. We finish this brief announcement by stating one result with mathematical precision.

An _incremental algorithm_ is an analytic endomorphism of the complex numbers of the form

$$I(z) = z + FR(h, f, z)$$

where $F = \dfrac{-f(z)}{f'(z)}$, $0 \le h \le 1$, f is a complex polynomial and $R(0, f, a) \equiv 0$. Thus, h, f could be thought of as parameters. The most famous example is the _incremental Newton method_ where $R \equiv h$, or simply Newton's method where $h = 1$ as well.

An important generalization is _incremental_ $k\underline{\text{th}}$ Euler's method $E_k(z) = z + FR$ where

$$R(h, f, z) = \frac{1}{F} T_k \left[\sum_{\ell=1}^{\infty} \frac{(f_z^{-1})^{(\ell)}}{\ell!} (f(z))(-hf(z))^{\ell} \right]$$

In this expression T_k is truncation at the $k\underline{\text{th}}$ power of h , and $(f_z^{-1})^{(\ell)}(f(z))$ is the $\ell\underline{\text{th}}$ derivative at $f(z)$ of the branch of f^{-1} which sends $f(z)$ to z . Here k is an integer between 1 and ∞ inclusive; E_1 is incremental Newton and

$$E_2(z) = z - \frac{f(z)}{f'(z)} (h - \sigma_2 h^2) , \text{ where } \sigma_2 = \frac{f^{(2)}(z)F}{2 f'(z)}$$

An _approximate zero_ of a complex polynomial f is defined as in the above reference (i.e., so that Newton's method (k = 1) converges rapidly). Let the space of initial points z_0 , be the set $S_R^1 = \{z \in \mathbb{C} | \ |z| = R\}$. Let $P_d(1)$ be the space of all complex polynomials $f(z) = \sum_{i=0}^{d} a_i z^i$, $a_d = 1$, $|a_i| \le 1$,

$i = 0 , \ldots , d - 1$. Take normalized Lebesgue measure on the product $S_R^1 \times P_d(1)$ for probability measure.

Theorem. There exist small universal positive constants K_1 , K_2 , K_3 , K_4 with the following true. Given d , μ , $0 < \mu < 1$, take k to be the least integer greater than log d ,

$$R = \frac{K_3}{\mu} \quad \text{and} \quad h = \frac{d \; \mu^{1+1/k} |\log \mu|}{K_4} \; .$$

Let $(z_0, f) \in S_R \times P_d(1)$. Then with probability $1 - \mu$, z_s will be an approximate zero of f where $z_{n+1} = E_k(z_n)$, $n = 0, 1, \ldots$, $s - 1$ and s is the least integer greater than

$$K_1 \; d \; \left[\frac{|\log \mu|}{\mu} \right]^{1+1/R} + K_2$$

Here μ is interpreted in terms of the normalized Lebesgue probability measure in $S_R^1 \times P_d(1)$.

ALGORITHMS FOR THE LINEAR COMPLEMENTARITY PROBLEM WHICH ALLOW AN

ARBITRARY STARTING POINT*

Dolf Talman[1] and Ludo Van der Heyden[2]

[1]Department of Econometrics, Tilburg University
Tilburg, The Netherlands
[2]School of Organization and Management
Yale University, New Haven, Connecticut, USA

1. Introduction

The linear complementarity problem with data $q \in R^n$ and $M \in R^{n \times n}$ consists in finding two vectory s and z in R^n such that

(1.1) $s = Mz + q$,

(1.2) $s, z \geq 0$,

(1.3) $s_i z_i = 0$, $i = 1, 2, \ldots, n$.

We denote this problem LCP or LCP(q,M). We only consider vectors (s,z) satisfying (1.1). Two vectors s and z are said to be feasible if they are nonnegative (1.2) and are said to be complementary if they satisfy (1.3).

The LCP is an important problem in mathematical programming [see, e.g., Garcia and Gould (1980) for references]. Lemke (1965) first presented a solution for this problem. His ideas were later

*The research in this paper was supported by the Office of Naval Research Contract Number N00014-77C-0518. We also are grateful to the referees for their helpful comments.

exploited by Scarf (1967) in his work on fixed point algorithms. The relationship between the LCP and the fixed point problem is well described by Eaves and Scarf (1976) and by Eaves and Lemke (1981).

Recently, Van der Laan and Talman (1979, 1981) proposed a class of variable dimension restart algorithms for approximating fixed points. These methods allow a start at an arbitrary point in the domain of the fixed point problem. One among several directions is followed to leave the starting point. These directions define a collection of cones of variable dimensions in which the search for an approximate fixed point takes place. Properties of the function govern the movement of the procedure between the conical regions. In each region movement occurs through simplicial pivoting, but continuous path-following could be applied too [see Allgower and Georg (1980)].

The intimate relation between the fixed point problem and the LCP raises the question of the significance of Van der Laan and Talman's work for the LCP. We show that the ideas behind their variable dimension fixed point algorithms yield an interesting class of LCP algorithms. An important feature of these algorithms is that they can be initialized at any nonnegative point z^0. When $z^0 = 0$, the algorithms reduce to Lemke's original algorithm (Lemke, 1965). Similar ideas can be used to modify other LCP algorithms, like the variable dimension algorithm of Van der Heyden (1980) [see also Yamamoto (1981)], to accept an arbitrary starting point. Flexibility in the choice of the starting point is desirable, e.g., in using prior information on the solution, in sensitivity analysis, and when solving nonlinear complementarity problems via a succession of approximating LCP's [Josephy (1979)].

Several authors have presented LCP algorithms which allow an arbitrary starting point. Eaves (1978) and Garcia and Gould (1980) present procedures based on homotopies. Reiser (1978), in an appendix to his dissertation, states two ways to transform an LCP with arbitrary starting point into one to which Lemke's algorithm can be applied. Our approach unifies the two Reiser algorithms in that the first Reiser algorithm becomes a special case in our framework, while another instance in our class of algorithms is very close to Reiser's second algorithm. This relationship with Reiser's work mirrors the similarity that exists between the Reiser and the Van der Laan and Talman fixed point algorithms [Reiser (1981)].

The paper is organized as follows. In section 2 we motivate our algorithm by interpreting the artificial variable in Lemke's algorithm as a measure of infeasibility. We then define the positions of our algorithm and the line segments which are followed to reach successive positions and which form a piecewise linear path leading to a solution. The procedure itself is explained in section 3, where we deal with convergence issues. In section 4, we discuss implementation and show that our algorithm can be seen as applying Lemke's algorithm to a transformed problem.

2. Movements and positions

We only consider pairs (s,z) satisfying (1.1) with z feasible. Let us take a starting point (s^0,z^0) . We define t_0 as

(2.1) $t_0 = \max (t_j : j \in I(n+k))$,

where, for any positive integer h , $I(h)$ denotes the index set $\{1,2,\ldots,h\}$, and where

$$t_i = -s_i \quad \text{for} \quad i \in I(n) ,$$

$$= \sum_{j \in P_h} s_j \quad \text{for} \quad i = n+h, \quad h \in I(k) ,$$

$\{P_h : h \in I(k)\}$ being an arbitrary partition of the set $I^+(n) = \{ i \in I(n) : z_i^0 > 0 \}$. The quantity t_0 measures the the infeasibility of the starting point (s^0,z^0) by checking for the nonnegativity of s^0 and for its complementarity with z^0 . z^0 is a solution for the LCP if and only if $t_0 \leq 0$. t_0 is negative at

(s^0, z^0) only if $(s^0, z^0) = (q,0)$ is a solution and $q > 0$. The largest infeasibility at z^0 defines the initial value of t_0.

Each component of the vector $t = (t_j : j \in I(n+k))$ is associated with a direction that can be followed to leave z^0. The directions associated with the first n components of t are the unit directions:

$$d^i = u^i \quad \text{for} \quad i \in I(n) \ ,$$

u^i denoting the ith unit vector in R^n. These also are the directions that can be followed to leave the starting point in Lemke's algorithm $(z^0 = 0)$. Leaving z^0 along d^i amounts to increasing z_i. With t_{n+h}, $h \in I(k)$, we associate direction d^{n+h} where

$$d_i^{n+h} = -z_i^0 \quad \text{for} \quad i \in P_h \ ,$$

$$= 0 \quad \text{for} \quad i \in I(n)-P_h \ .$$

A movement along d^{n+h} amounts to decreasing all coordinates of z with indices in P_h. The directions are illustrated in figure 1.

Figure 1 shows that the directions $D = (d^i : i \in I(n+k))$, when drawn through z^0, partition R^n into relatively open conical regions $C(P) = \{z : z = z^0 + Dy \ , \ y \in R^{n+k} \ , \ y_j > 0 \text{ for } j \in P\}$. To maintain the feasibility of z we require that

$$(2.2) \qquad y_j \leq 1 \quad \text{when} \quad j > n \ .$$

A vector y is said to be **feasible** if it is nonnegative, satisfies (2.2), but does not meet $y_j > 0$ for all $j \in P_h u \{n+h\}$, $h \in I(k)$. The latter condition ensures that the correspondence between y and z is one-to-one. In what follows, we equivalently refer to z or to its unique representation in terms of a feasible y.

The algorithm maintains a generalized form of complementarity between leading infeasibilities in maximand (2.1) and directions. Except for boundary issues, $\underline{t_0\text{-complementarity}}$ between the vectors

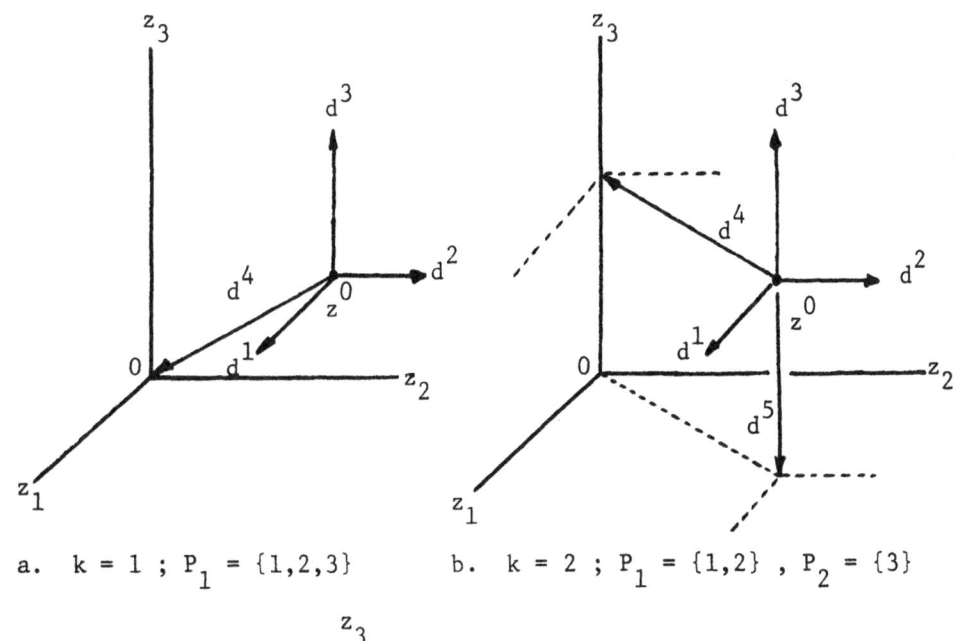

a. $k = 1$; $P_1 = \{1,2,3\}$ b. $k = 2$; $P_1 = \{1,2\}$, $P_2 = \{3\}$

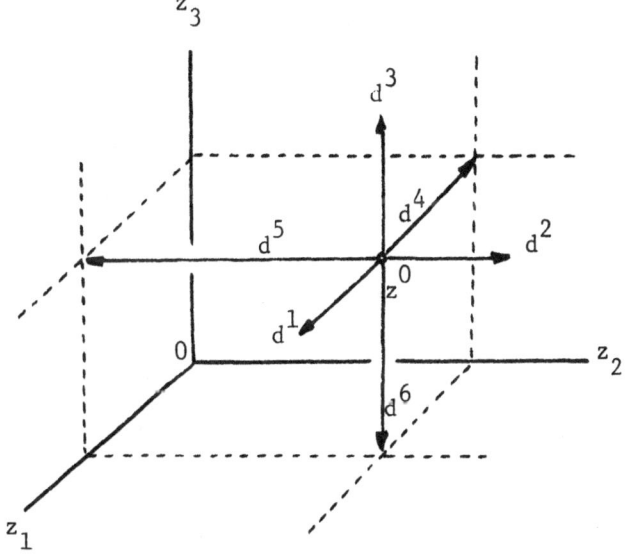

c. $k = 3$; $P_1 = \{1\}$, $P_2 = \{2\}$, $P_3 = \{3\}$

FIGURE 1. The directions d^j , $1 \le j \le n{+}k$, in three special
cases of our algorithm ($n = 3$; $k = 1$, 2, and 3).

t and y requires

$$y_j = 0 \quad \text{or} \quad t_j = t_0, \quad j \in I(n+k).$$

The algorithm thus moves in cones defined by directions associated with leading infeasibilities. We now motivate the definition of t_0-complementarity on the boundary of the nonnegative orthant in z-space.

Assume that $t_0 = t_{n+h}$ is the unique leading infeasibility at z^0. The only movement allowed by t_0-complementarity is to move along direction d^{n+h} by increasing y_{n+h}. As soon as in this movement another infeasibility, say t_j, ties t_{n+h} as the leading infeasibility $(t_0 = t_{n+h} = t_j)$, further movement along d^{n+h} is infeasible for it would lead to points (t,y) verifying $t_j = t_0 > t_{n+h}$ with $y_{n+h} > 0$. These points would not be t_0-complementary. The only movement consistent with t_0-complementarity is to then move in t_0-complementary fashion into cone $C(\{n+h,j\})$ by increasing y_j while maintaining $t_0 = t_{n+h} = t_j$. The latter restriction removes the degree of freedom introduced by moving into a higher dimensional region. Another possibility arising when leaving z^0 along d^{n+h} is that t_{n+h} remains the unique leading infeasibility so that, for all $0 \leq y_{n+h} \leq 1$ and for all $j \in I(n+k)-\{n+h\}$, $t_0 = t_{n+h} > t_j$. Once the boundary $y_{n+h} = 1$ is reached, further movement along d^{n+h} generates infeasible y's. The algorithm then keeps $y_{n+h} = 1$ and allows t_{n+h} to differ from t_0 by removing t_{n+h} from maximand (2.1). The definition of t_0 on the boundary of the feasible y-region is then completed as follows:

(2.3) $t_0 = \max(t_j : j \in I(n+k), y_j < 1 \text{ when } j > n)$.

In order to maintain t_0-complementarity during the movement of the algorithm, we need to generalize the notion of t_0-complementarity by also calling the pair (t_{n+h}, y_{n+h}) t_0-complementary when $y_{n+h} = 1$.

(2.4) <u>Definitions</u>. A component t_j is said to be <u>nonbasic</u> if

$t_j = t_0$. t_0 is said to be nonbasic when $t_0 = 0$. y_j is nonbasic when $y_j = 0$ or when $y_j = 1$ and $j > n$. The vectors t
and y are said to be t_0-complementary when for each $j \in I(n+k)$
either y_j or t_j is nonbasic.

t_0-complementarity is one of two properties which will be
shown to define a piecewise linear path to a solution. The second
property constrains the components of t which are not involved
in the computation of t_0 , namely those associated with components
of y assuming their upper bound. We motivate the second property
by returning to a situation discussed earlier. Let us imagine that
the algorithm leaves the initial point z^0 along direction d^{n+h}
and that this movement is pursued until $y_{n+h} = 1$. During this
movement t_0-complementarity requires that t_{n+h} remains the
largest component of t . Upon reaching the boundary, t_{n+h} dis-
appears from maximand (2.3) and t_0 decreases discontinuously if
the second largest component of t is strictly smaller than t_{n+h} .
Assuming this to be the case, t then verifies $t_{n+h} > t_0 = t_j$,
$j \in I(n+k) - \{n+h\}$. If we like the algorithm to terminate with a
solution when $t_0 = 0$, we must require that the inequality
$t_{n+h} > t_0$ be maintained while $y_{n+h} = 1$. If at a later stage
t_{n+h} again becomes equal to t_0 , then the algorithm continues in
t_0-complementary fashion by decreasing y_{n+h} from 1 while maintaining $t_{n+h} = t_0$. We now formally introduce the lines followed
by the algorithm.

(2.5) <u>Definition</u>. A <u>line of our algorithm</u> consists of a set of
t_0-complementary points such that

    a.   exactly one variable in each pair (t_j, y_j) is nonbasic;

    b. $t_j > t_0$ when $y_j = 1$, $j > n$;

    c. $t_0 > 0$.

Note that by definition of t_0 , $t_j \leq t_0$ when $y_j = 0$, while the algorithm requires that $t_j \geq t_0$ when $y_j = 1$, $j > n$. In all other cases $(0 < y_j$ and $y_j < 1$ when $j > n)$, $t_j = t_0$. The algorithm thus imposes various types of constraints on the t-variables in different regions of z-space. Figure 2 illustrates these constraints for the case $n = 2$ and $z^0 > 0$.

In order for the set of points satisfying (2.5) to form a line, we need to impose the following nondegeneracy assumption.

(2.6) Assumption. At most $n+k+1$ among the $2(n+k)+1$ variables (t_0, t, y) are nonbasic at any given point (t,y).

We indicate in section 4 that this assumption is similar to a nondegeneracy assumption in linear programming and thus can be satisfied with the usual perturbation techniques. One component of t is nonbasic by definition of t_0. t_0-complementarity imposes $n+k-1$ additional restrictions on the vector (t,y) so that one degree of freedom remains. The set of points (if any) satisfying definition (2.5) with a fixed set of nonbasic variables do form a line segment.

Let us examine the endpoints of the lines of our algorithm. An endpoint is reached when a basic variable becomes nonbasic. If there is no discontinuity in the value of t_0 and if t_0 is still basic, there is by nondegeneracy exactly one pair of variables which are both nonbasic. This gives rise to two types of position for the algorithm. At a _position of type a_ we have that, for some $j \in I(n+k)$, $y_j = 0$ and $t_j = t_0 > 0$. At a _position of type b_ we have that, for some $j > n$, $y_j = 1$ and $t_j = t_0 > 0$. If an endpoint is reached where t_0 is nonbasic, then it will be shown to be a solution. The latter is also true if t_0 becomes nonpositive during a discontinuous decrease at the endpoint. If after a discontinuous decrease t_0 is still positive, there is one nonbasic pair (y_h, t_h) with $t_h = t_0 > 0$ and $y_h = 0$ for some $h \in I(n+k)$. The endpoint is a position of type a. This completes our

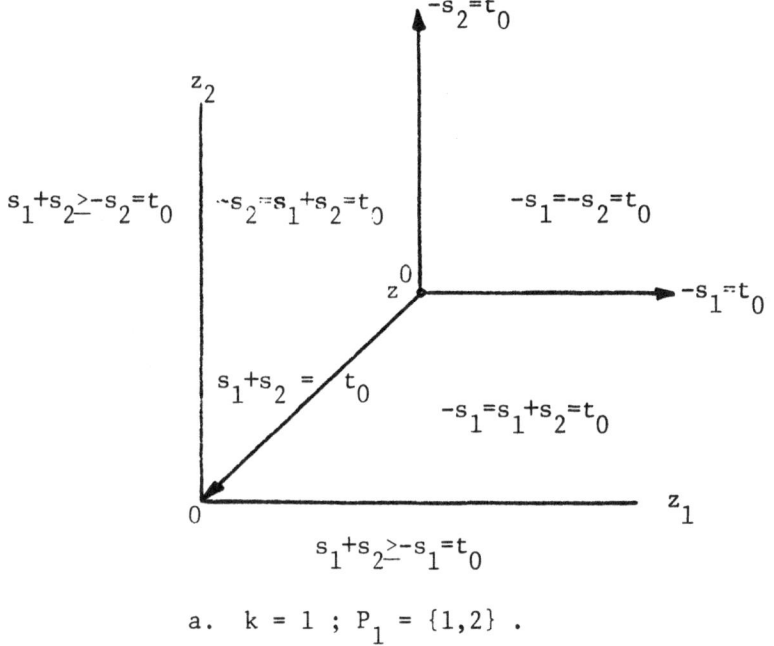

a. $k = 1$; $P_1 = \{1,2\}$.

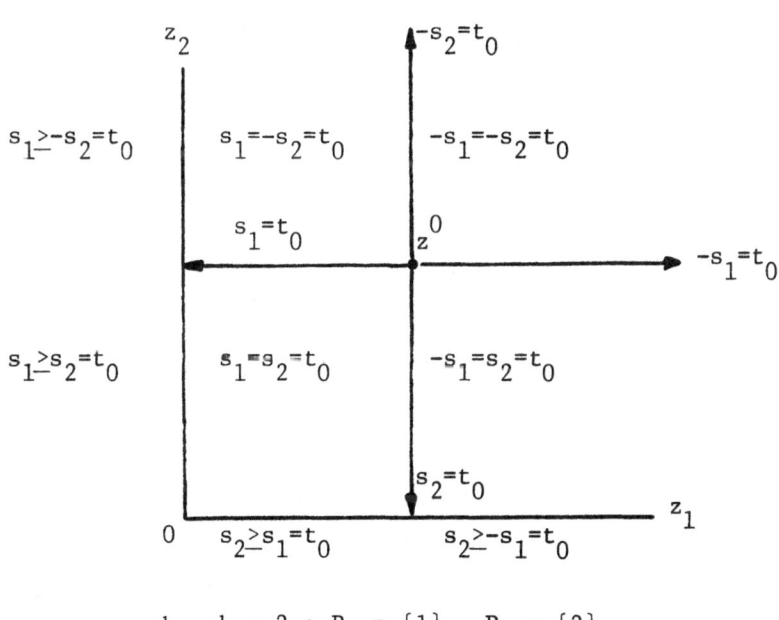

b. $k = 2$; $P_1 = \{1\}$, $P_2 = \{2\}$.

FIGURE 2. The constraints imposed on the variables in different regions of z-space for a 2-dimensional example (n=2). We have omitted the inequalities that are implicit in the definition of t_0 .

classification of endpoints into positions of type   a   or   b   and

t_0-complementary points with $t_0 \leq 0$.

   We now prove the important fact that if t_0 becomes nonpos-

itive at an endpoint, then a solution has been found. Since

$t_0 \geq \max(t_i = -s_i : i \ \varepsilon \ I(n))$, it is clear that $s \geq 0$ at such an

endpoint. We still need to argue that $s_i = 0$ whenever $z_i > 0$.

We distinguish two cases. If $i \ \varepsilon \ P_h$ and $y_{n+h} < 1$, then $s_i = 0$

follows from the fact that $0 \geq t_0 \geq t_{n+h} = \sum_{i \varepsilon P_h} s_i \geq 0$. If

$y_{n+h} = 1$, then the positivity of z_i requires the positivity of

y_i along the line leading to the endpoint. Hence, t_i is nonbasic

along the line: $-s_i = t_i \geq 0$ (since $t_0 > 0$ along the line).

This inequality is still valid at the endpoint and implies $s_i = 0$.

   We illustrate the incidence between positions and lines of our

algorithm in Figure 3. The algorithm leaves the initial position

along the unique line incident to it. Every other position, which

is not a solution, has two lines incident to it. If the position

is reached along one line, then the algorithm leaves it along the

other line. Solutions can be shown to be incident to only one line

of our algorithm.

3.   Convergence issues

   The previous section set the stage for an application of the

well-known Lemke-Howson argument. The initial position is incident

to one line of the algorithm. Every other position which is not a

solution is incident to two lines of our algorithm. The Lemke-

Howson argument proves that if lines are followed without turning

back no position will ever be visited twice. The number of lines

is finite, hence, so is the number of positions. The algorithm

thus either stops at a solution for the LCP or follows an unbounded

line. Following Lemke (1965), we present a class of matrices--

characterized by Garcia (1973)--for which the algorithm finds a

solution for any right-hand side vector q . We then show that for

copositive plus matrices [Lemke (1965)] the existence of an un-

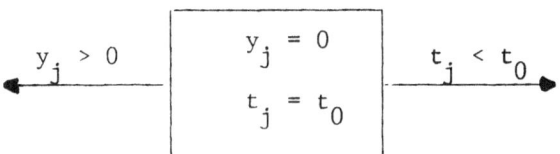

i.   Position of type a: at least one y-variable is basic. No
     discontinuity in value of t_0 .

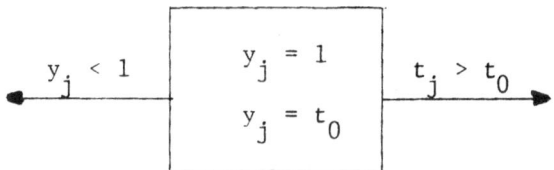

ii. Position of type a: all y-variables are nonbasic. Let
     $n+h = \text{argmin}(t_{n+j} : y_{n+j} = 1)$. Discontinuous increase in
     value of t_0 when leaving position along line drawn at the
     right of the position.

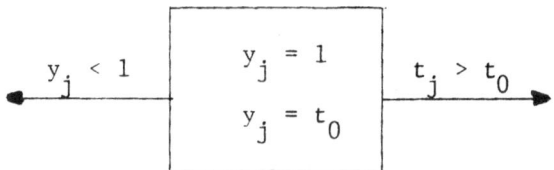

iii. Position of type b: $j > n$. No discontinuity in value of t_0 .

FIGURE 3. The incidence between positions (with $t_0 > 0$) and lines
of our algorithm. Notice that in case ii, the line drawn at the
right of the position is defined only if $\{n+i : y_{n+i} = 1\}$ is non-
empty. If the set is empty, we are at the initial position. This
position is the only position (with $t_0 > 0$) incident to one line
of our algorithm.

bounded line implies that the LCP is not feasible. The point
behind both results is that they hold for any starting point z^0
in R_+^n. [Garcia and Gould (1980) discuss the possibility of convergence for a particular set of starting points.]

(3.1) Theorem. Let M satisfy the property that LCP(q,M) admits
the unique solution $z = 0$ both when $q = 0$ and when $q = e$,
where $e = (1,1,...1)^t$. Then no line of our algorithm is unbounded.

Proof. An unbounded line of our algorithm implies the existence of
a $(2n+1)$-directional vector $(\bar{t}_0, \bar{s}, \bar{z})$ verifying the following
conditions:

(3.2)   a. $\bar{s} = M\bar{z}$ with $\bar{z} \geq 0$;

           b. if $\bar{z}_i > 0$ then $-\bar{s}_i = \bar{t}_0$;

           c. if $\bar{z}_i = 0$ then $-\bar{s}_i \leq \bar{t}_0$;

           d. $\bar{t}_0 \geq 0$.

[Notice that the directional vector $\bar{y}$ associated with $\bar{z}$ always
has $\bar{y}_j = 0$ for $j > n$, for we can't leave the nonnegative orthant
in z-space. Hence, $\bar{y}_i = \bar{z}_i$ for $i \in I(n)$.] It is clear that $\bar{z}$
is nonzero. If $\bar{t}_0 = 0$ then $\bar{s}$ is nonnegative and complementary
with $\bar{z}$, which itself is nonnegative. $\bar{z}$ represents a nontrivial
solution for LCP($0,M$), which is impossible. If $\bar{t}_0 > 0$, we rescale
$\bar{s}$ and $\bar{z}$ so that $\bar{t}_0 = 1$. $\bar{z}$ satisfies the inequalities
$M\bar{z} + e \geq 0$, where the ith inequality is an equality if $\bar{z}_i > 0$.
LCP(e,M) thus admits a nonzero solution, which again contradicts
our assumption.

(3.3) Theorem. Let M be copositive plus: $u^t M u \geq 0$ when $u \geq 0$,
with $u^t M u = 0$ implying $(M+M^t)u = 0$. If the algorithm generates
an unbounded line then the LCP is infeasible.

Proof. The LCP is infeasible if $s = Mz + q$, s and $z \geq 0$, is an infeasible linear system. By Farkas's lemma this infeasibility is equivalent with the existence of a nonnegative vector u such that $u^t M \leq 0$ and $u^t q < 0$.

The arguments of Theorem (3.1) show that an unbounded line implies the existence of a vector $(\bar{t}_0, \bar{s}, \bar{z})$ verifying (3.1). If $\bar{t}_0 > 0$, then $\bar{z}^t M\bar{z} = \bar{z}^t\bar{s} = -(\bar{z}^t e)\bar{t}_0 < 0$ since $\bar{z}$ is nonzero. This contradicts the copositive plus character of M. Hence $\bar{t}_0 = 0$.

A zero value for $\bar{t}_0$ implies that $\bar{z}^t M\bar{z} = 0$ and, hence, $M^t\bar{z} = -M\bar{z} \leq 0$, since $-M\bar{z} = -\bar{s} \leq \bar{t}_0 e = 0$. $\bar{z}$ is our candidate Farkas direction. To conclude our proof, we only need to show that $\bar{z}^t q < 0$.

Consider the unique endpoint of the unbounded line, say (t_0^*, s^*, z^*), where

$$s^* = Mz^* + q, \qquad z^* \geq 0 \text{ and } t_0^* > 0.$$

Premultiplication with $\bar{z}^t$ yields

$$\bar{z}^t s^* = \bar{z}^t M z^* + \bar{z}^t q$$

$$= -\bar{s}^t z^* + \bar{z}^t q.$$

Because of t_0-complementarity at (s^*, z^*) and along the unbounded line we have $-s_i^* = t_0^*$ whenever $\bar{z}_i > 0$, for even if nonbasic at z^*, y_i is basic along the line. Hence, $\bar{z}^t s^* = -(\bar{z}^t e)t_0^* < 0$ implying that

$$-\bar{s}^t z^* + \bar{z}^t q < 0.$$

If we can argue that $\bar{s}^t z^* = 0$, then our result is obtained.

If $\bar{s}_i > 0 = -\bar{t}_0$, then y_i is nonbasic along the unbounded line $(y_i = z_i = 0)$. At the same time,

$$\sum_{j \in P_h} \bar{s}_j \geq \bar{s}_i > 0 = \bar{t}_0,$$

where $i \in P_h$. The first inequality follows from the nonnegativity

of $\bar{s}$. Inequality $\sum_{j \in P_h} \bar{s}_h > \bar{t}_0$ implies that $y_{n+h} = 1$ along the

unbounded line, and thus at its endpoint z^* . Since $y_{n+h} = 1$

and $y_i = 0$ along the line, we have $z_i = 0$ along the line, and

hence $z_i^* = 0$ at the endpoint. This concludes the argument

establishing that $\bar{s}^t z^* = 0$.

4. Implementation

We introduce the matrix $E = (E_{hj})$ to identify the partition
$\{P_h : h \in I(k)\}$:

$$E_{hj} = 1 \quad \text{if} \quad j \in P_h ,$$

$$= 0 \quad \text{otherwise} .$$

t can then be written in matrix form as

$$(4.1) \qquad t = \begin{bmatrix} -s \\ \\ Es \end{bmatrix} = \begin{bmatrix} -M \\ \\ EM \end{bmatrix} z + \begin{bmatrix} -q \\ \\ Eq \end{bmatrix} .$$

We introduce nonnegative vectors to represent the deviations of t

from t_0:

$$t = \begin{bmatrix} e^1 \\ \\ e^2 \end{bmatrix} t_0 - \begin{bmatrix} t^{1-} \\ \\ t^{2-} \end{bmatrix} + \begin{bmatrix} 0 \\ \\ t^{2+} \end{bmatrix} ,$$

e^1 and $t^{1-} \in R_+^n$, e^2 , t^{2-} and $t^{2+} \in R_+^k$, e^i $(i=1,2)$ re-
presenting a vector of ones. We partition $y = (y^1, y^2)^t$,
$y^1 \in R_+^n$, $y^2 \in R_+^k$, and introduce the corresponding partition
for $D = (D^1, D^2)$. We write the feasibility constraint on y^2 as

$$v^2 + y^2 = e^2 , \quad v^2 \text{ and } v^2 \geq 0 ,$$

and append it to (4.1) . The latter system can be written

$$(4.2) \quad \begin{bmatrix} t^{1-} \\ t^{2-} \\ v^2 \end{bmatrix} = \begin{bmatrix} MD^1 & MD^2 & 0 \\ -EMD^1 & -EMD^2 & I \\ 0 & -I & 0 \end{bmatrix} \begin{bmatrix} y^1 \\ y^2 \\ t^{2+} \end{bmatrix} + \begin{bmatrix} e^1 \\ e^2 \\ 0 \end{bmatrix} t_0 + \begin{bmatrix} q^0 \\ -Eq^0 \\ e^2 \end{bmatrix} ,$$

where $q^0 = Mz^0 + q$. t_0-complementarity between t and y is equivalent with the ordinary complementarity between (t^{1-}, t^{2-}, v^2) and (y^1, y^2, t^{2+}) . The starting point is $(y^1, y^2, t^{2+}) = 0$. Our algorithm can thus be seen as a projection of Lemke's algorithm applied to an enlarged problem. Notice also that assumption (2.6) is satisfied when linear system (4.2) is nondegenerate. Classical perturbation techniques applied to (4.2) ensure nondegeneracy. Finally, the discontinuity of t_0 , as described in figure 3 (case ii), reduces to a trivial pivot step in the enlarged system. In the pivot step that corresponds in figure 3 (case ii) to a movement along the line appearing at the right-hand side of the position, t_0 increases by an amount equal to the samllest positive component of t^{2+} . All basic components of t^{2+} are decreased by that amount whereas all basic components of (t^{1-}, t^{2-}) are increased by a similar amount. The components of (y^1, y^2) are not affected by this pivot step.

It is clear that the last k equations in (4.2) can be handled implicity as they represent upper bounds on y^2 . We now indicate that a similar implicit treatment can be given to the middle k equations. Adding appropriate sums of the first n equations to these middle k equations, they can be written

$$(4.3) \qquad Et^{1-} + t^{2-} - t^{2+} - (Ee^1 + e^2) t_0 = 0 .$$

These equations are of the GVUB type [Schrage (1978)] since every

variable with a positive coefficient appears only once in (4.3) .
At a position $t_0 > 0$ and $t^{2+} \geq 0$, so that at least one among

the variables t_h^{2-} and $(t_j^{1-} : j \in P_h)$ is basic. This implies
that the basis matrix, after suitable permutation of its columns,
contains an identity submatrix of order k . This property allows
an implicit treatment of these equations so that every pivot step
in system (4.2) involves the updating of a basic submatrix of
order n , rather than $n+2k$ in an explicit treatment of (4.2) .

   There may exist instances of the LCP where the freedom to
arbitrarily choose a partition of $I^+(n)$ could be exploited. One
such instance occurs when the matrix M presents the special
structure

$$M = \begin{bmatrix} A, & 0 & & & 0 \\ 0 & A_2 & \cdots & & 0 \\ & & \ddots & & \\ \vdots & & & \ddots & \\ B_1 & B_2 & \cdots & & A_k \end{bmatrix} .$$

Every submatrix A_h could then be associated with a partition set
P_h . However, in the absence of special structure, it is reasonable
to expect the algorithm to treat all coordinates symmetrically.
This points us to the two extreme cases, $k = 1$ and $k = \left| I^+(n) \right|$.
   When $k = \left| I^+(n) \right|$, every set P_h is a singleton. If $z^0 > 0$
(4.2) can be written

$$(4.4) \qquad \begin{bmatrix} t^{1-} \\ t^{2-} \\ v^2 \end{bmatrix} = \begin{bmatrix} M & -M & 0 \\ -M & M & I \\ 0 & -I & 0 \end{bmatrix} \begin{bmatrix} y^1 \\ y^2 \\ t^{2+} \end{bmatrix} + \begin{bmatrix} e \\ e \\ 0 \end{bmatrix} t_0 + \begin{bmatrix} q^0 \\ -q^0 \\ z^0 \end{bmatrix} .$$

If $z_i^0 = 0$, feasibility in row $2n+i$ sets $v_i^2 = y_i^2 = 0$, so that
equation $n+i$ can be eliminated, as indicated in (4.2) . This

case is analogous to one of Reiser's algorithms [Reiser (1978)].

When $k = 1$, (4.2) becomes

$$(4.5) \quad \begin{bmatrix} t^{1-} \\ t^{2-} \\ u^2 \end{bmatrix} = \begin{bmatrix} M & -Mz^0 & 0 \\ -e^+M & e^+Mz^0 & 1 \\ 0 & 1 & 0 \end{bmatrix} \begin{bmatrix} y^1 \\ y^2 \\ t^{2+} \end{bmatrix} + \begin{bmatrix} e \\ 1 \\ 0 \end{bmatrix} t_0 + \begin{bmatrix} q^0 \\ -e^+q^0 \\ 1 \end{bmatrix} ,$$

where t^{2-}, t^{2+}, u^2, and y^2 are scalars, and where $e^+ = (e_i^+)$
with $e_i^+ = 1$ if $z_i^0 > 0$, and $e_i^+ = 0$ otherwise. The second
Reiser algorithm considers only the first n equations of (4.5).
That algorithm corresponds to movements along t_0-complementary
lines where $t_0 = \max(-s_1, -s_2, \dots, -s_n, 0)$ as compared with
$t_0 = \max(-s_1, -s_2, \dots, -s_n \cdot \sum_{i \in I^+(n)} s_i)$ for our algorithm. The comple-

mentarity conditions along a line in Reiser's algorithm are

$$(4.6) \quad (t^{1-})^t y^1 = 0 \quad \text{and} \quad t_0 y^2 = 0 .$$

In this setting $t_0 = 0$ no longer identifies a solution. The
algorithm termimates either when y^2 reaches its upper bound of 1
or when t_0 and t^{1-} are all nonbasic. In the first case, $t_0 = 0$
by complementarity along a line so that the first n equations of
(4.5) can be written $t^{1-} = My^1 + q$. Since (t^{1-}, y^1) is also
complementary it is a solution for the LCP. In the second case,
$(t_0, t^{1-}) = 0$ and it is easily seen that $(s,z) = (0, y^1 + (1-y^2)z^0)$
is a solution for the LCP.

We conclude with examining the special case where $z^0 = 0$.
Equation (4.2) then becomes

$$t^{1-} = My^1 + e^1 t_0 + q .$$

Our algorithm requires t^{1-} and y^1 to remain complementary and
this special case thus reduces to Lemke's original algorithm.

REFERENCES

Allgower, E. L. and K. Georg (1980), "Simplicial and continuation methods for approximating fixed points and solutions to systems of equations," SIAM Review, 22, pp. 28-85.

Eaves, B. C. (1978), "Computing stationary points," Mathematical Programming Study, 7, pp. 1-14.

Eaves, B. C. and C. E. Lemke (1981), "Equivalence of LCP and PLS," Mathematics of Operations Research, 6, pp. 475-484.

Eaves, B. C. and H. Scarf (1976), "The solution of systems of piecewise linear equations," Mathematics of Operations Research, 1, pp. 1-27.

Garcia, C. B. (1973), "Some classes of matrices in linear complementarity theory." Mathematical Programming, 5, pp. 299-310.

_____ and F. J. Gould (1980), "Studies in linear complementarity," Center for Mathematical Studies in Business and Economics, University of Chicago, Chicago.

Josephy, N. (1979), "Newton's method for generalized equations," Technical Summary Report #1965, Mathematics Research Center, University of Wisconsin, Madison.

Van der Laan, G. and A. J. J. Talman (1979), "A restart algorithm for computing fixed points without an extra dimension," Mathematical Programming, 17, pp. 74-84.

_____ (1981), "A class of simplicial restart fixed point algorithms without an extra dimension," Mathematical Programming, 20, pp. 33-48.

Lemke, C. E. (1965), "Bimatrix equilibrium points and mathematical programming," Management Science, 11, pp. 681-689.

Reiser, P. M. (1978), "Ein hybrides Verfahren zur Lösung von nichtlinearen Komplementaritäts-problemen und seine Konvergenz-eigenschaften," Dissertation, Eidgenössischen Technischen Hochschule, Zurich, Switzerland.

_____ (1981), "A modified integer labeling for complementarity algorithms," Mathematics of Operations Research, 6, pp. 129-139.

Scarf, H. (1967), "The approximation of fixed points of a continuous mapping," SIAM Journal on Applied Mathematics, 15, pp. 1328-1342.

REFERENCES

Schrage, L. (1978), "Implicit representation of generalized upper bounds in linear programming," Mathematical Programming, 14, pp. 11-20.

Van der Heyden, L. (1980), "A variable dimension algorithm for the linear complementarity problem," Mathematical Programming, 19, pp. 328-346.

Yamamoto, Y. (1981), "A note on Van der Heyden's variable dimension algorithm for the linear complementarity problem," Discussion Paper No. 103, Institute for Socio-Economic Planning, University of Tsukuba, Ibaraki, Japan.

ENGINEERING APPLICATIONS OF THE CHOW-YORKE ALGORITHM

Layne T. Watson

Department of Computer Science
Virginia Polytechnic Institute & State University
Blacksburg, Virginia 24061 USA

Abstract. The Chow-Yorke algorithm is a scheme for developing
homotopy methods that are globally convergent with probability one.
Homotopy maps leading to globally convergent algorithms have been
created for Brouwer fixed point problems, certain classes of nonlinear systems of equations, the nonlinear complementarity problem,
some nonlinear two point boundary value problems, and convex optimization problems. The Chow-Yorke algorithm has been successfully
applied to a wide range of engineering problems, particularly those
for which quasi-Newton and locally convergent iterative techniques
are inadequate. Some of those engineering applications are surveyed
here.

1. Why homotopy methods?

A frequently asked and legitimate question is "Why do you need a
homotopy method?" Just because a homotopy method is theoretically
elegant and can be proven globally convergent does not justify its
use if a simpler and more efficient method would suffice. The intent
of this paper is to present a list of problems for which Newton and
quasi-Newton methods are either totally inadequate or much more expensive than a globally convergent homotopy method. Consider the
example

$$t^2 u - 1 = 0$$

$$u^2 - 1 = 0 \ .$$

This is a one-dimensional case of a structural design problem where
t is the material thickness and u is the displacement. For this

problem, Newton's method started from $(t,u) = (-2,-2)$ diverges.
Very robust, well programmed quasi-Newton methods also fail. For
example, least change secant update algorithms (sometimes erroneously
called globally convergent), started at $(0,-1)$ fail because $(0,-1)$
is a local minimum for the norm of the function. This local minimum phenomenon is typical of fluid dynamics and elastica problems.
Let

$$x = \begin{pmatrix} t \\ u \end{pmatrix}, \quad f(x) = \begin{pmatrix} t^2 u - 1 \\ u^2 - 1 \end{pmatrix}, \quad \text{and}$$

$$\rho(\lambda,x) = \lambda f(x) + (1-\lambda)(x-c).$$

Using the latter homotopy is also unsuccessful since the zero curve
of $\rho(\lambda,x)$ does not reach $\lambda = 1$ (see Figure 1).

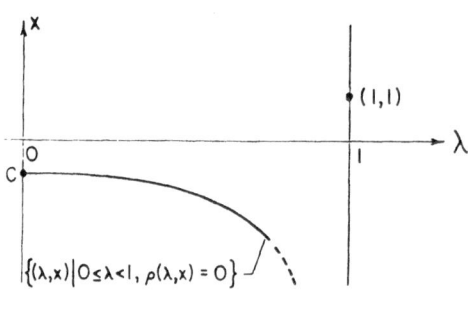

Figure 1

However, the homotopy map

$$\rho(\lambda,x) = f(x) - (1-\lambda) \begin{pmatrix} a \\ b \end{pmatrix}$$

does work. It is possible to prove that for almost all $\begin{pmatrix} a \\ b \end{pmatrix} \in E^1 \times$
$(0,1)$ zero curves of $\rho(\lambda,x)$ reaching a solution exist [57]. See
Figure 2. This example shows that there is probably not a "homotopy
map for all seasons", but that some homotopy map, resulting in a
globally convergent algorithm, may exist.

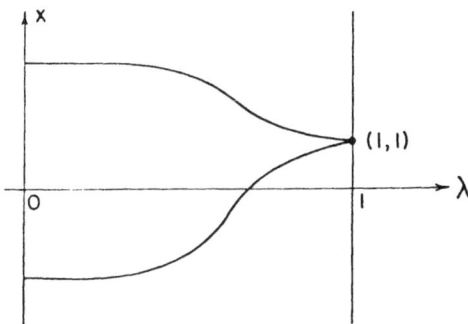

Figure 2.   Zero curves of $\rho(\lambda,x)$ for different
parameters $\binom{a}{b}$.

2.   The Chow-Yorke Algorithm

The theoretical foundation of the Chow-Yorke algorithm is given
in the following lemma [9,10,51]:

Def.   Let   U, V $\subset$ E^n be open sets and $\rho : U \times (0,1) \times V \rightarrow E^n$
be a C^2 map. ρ is said to be transversal to zero if the Jacobian
matrix   Dρ has full rank on $\rho^{-1}(0)$.

Parameterized Sard's Theorem.   If $\rho(a,\lambda,x)$ is transversal to zero,
then for almost all   a ε U   the map

$$\rho_a(\lambda,x) = \rho(a,\lambda,x)$$

is also transversal to zero; i.e., with probability one the Jacobian
matrix $D\rho_a(\lambda,x)$ has full rank on $\rho_a^{-1}(0)$.

The geometric interpretation of this result is that the set

$$\rho_a^{-1}(0) = \{(\lambda,x)\,|\,0 \le \lambda < 1,\ \rho_a(\lambda,x) = 0\}$$

of zeros of ρ_a consists of smooth, disjoint curves which have no
endpoints in $(0,1) \times V$ and have finite arc length in any compact
subset of $(0,1) \times V$. This holds for almost all a, or, in other
words, with probability one. See Figure 3.

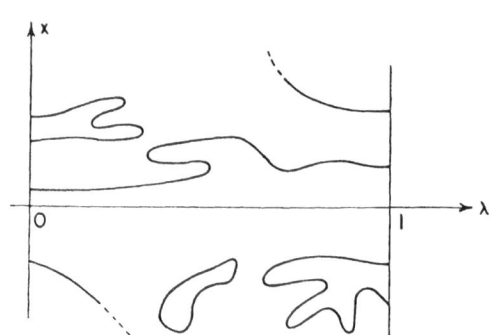

Figure 3. Typical zero set of $\rho_a(\lambda,x)$.

The recipe for a globally convergent algorithm is then:

1) Construct a homotopy map $\rho(a,\lambda,x)$ such that

   a) ρ is transversal to zero;

   b) $\rho_a(0,x) = 0$ is trivial to solve, and preferably has a uni-
      que solution;

   c) $\rho_a(1,x) = 0$ is equivalent to the given problem.

2) Prove that the zero curves of ρ_a emanating from $\lambda = 0$ are
   bounded (and monotone in λ if $\rho_a(0,x) = 0$ has more than one
   solution).

If 1) and 2) above have been accomplished, then for almost all a
there exists a zero curve ν of ρ_a, along which the Jacobian
matrix $D\rho_a(\lambda,x)$ has full rank, emanating from $\lambda = 0$ and reaching
a solution of the given problem at $\lambda = 1$ [10,51]. Thus a globally

convergent algorithm consists of tracking this zero curve ν of
ρ_a from $\lambda = 0$ until it reaches $\lambda = 1$.   The "Chow-Yorke algorithm"
refers to 1), 2), and any scheme for tracking this zero curve ν of
the homotopy map $\rho_a(\lambda, x)$.

There is some controversy over how to track this zero curve ν.   A
scheme is summarized here (see [51,52] for more details) which the
author has found to be accurate, easy to use, reliable, robust, and
efficient for practical problems.   Since the zero curve ν is smooth,
it can be parameterized by arc length   s.   Thus, $\lambda = \lambda(s)$, $x = x(s)$
along ν and

$$\rho_a(\lambda(s), x(s)) = 0 \tag{1}$$

identically in   s.   Let ν emanate from $(0, x_0)$.   Then ν is the
trajectory of the initial value problem

$$\frac{d}{ds} \rho_a(\lambda(s), x(s)) = D\rho_a(\lambda(s), x(s)) \begin{pmatrix} \dfrac{d\lambda}{ds} \\[2mm] \dfrac{dx}{ds} \end{pmatrix} = 0 , \tag{2}$$

$$\left\| \begin{pmatrix} \dfrac{d\lambda}{ds} \\[2mm] \dfrac{dx}{ds} \end{pmatrix} \right\|_2 = 1 \quad , \tag{3}$$

$$\lambda(0) = 0, \ x(0) = x_0 \tag{4}$$

Recall that (for almost all   a) the Jacobian matrix

$$D\rho_a(\lambda(s), x(s)) \tag{5}$$

has full rank.   Therefore, (5) has a one-dimensional kernel, and
$(d\lambda/ds, dx/ds)$ is uniquely determined by (2), (3), and continuity.
The kernel of the matrix (5) is determined in a numerically stable
and accurate way by factoring (5) with Householder reflections [7,
51, 56].   Values of $(d\lambda/ds, dx/ds)$ are used as input to an ODE
solver which solves the initial value problem (2-4).   Since evalua-

tion and factorization of the Jacobian matrix (5) is expensive, an
ODE solver which puts a premium on minimizing the number of derivative evaluations seems appropriate. For example, the subroutines
STEP and INTRP of [42] work very well in this context. For some
practical considerations regarding the tracking of ν and obtaining
the solution at λ = 1, see [51], [52], [55], and [56].

3. Engineering Applications

To give some idea of how widely applicable the Chow-Yorke algorithm
is, a partial list of problems solved by the Chow-Yorke algorithm
is presented. These problems range from fairly simple to extremely
difficult, and Newton-type methods either partially or totally failed
on all of them.

1. Elliptic porous slider.

2. Squeezing of a viscous fluid between parallel plates.

3. Squeezing of a viscous fluid between elliptic plates.

4. Viscous flow between rotating discs with injection on the porous
disc.

5. Deceleration of a rotating disc in a viscous fluid.

6. Porous channel flow in a rotating system.

7. Optimal structural design (continuum mechanics).

8. Convex unconstrained optimization.

9. Optimization with nonnegativity constraints.

10. Nonlinear complementarity problem.

11. Large deformation of an elastic rod.

12. Large deformation of C-clamps.

13. Large deformation of negator clips.

14. Fluid-filled cylindrical membrane container.

15. Circular leaf spring.

16. Hanging elastic ring.

17. Equilibrium of heavy elastic cylindrical shells.

18. Equilibrium of reticulated shells.

19. Collapse of tethered blood vessels.

A few of these will now be discussed in more detail.

ELLIPTIC POROUS SLIDER

Consider an air-cushioned vehicle, supported by air-pressure from air forced down through its base, with an elliptic base. The important quantities are lift, drag, and the most efficient direction in which to move the vehicle. The fluid flow is described by the nondimensional equations [54]:

$$R[(h')^2 - (h + k)h''] \;=\; Q + h'''$$

$$R[(k')^2 - (h + k)k''] \;=\; \beta^2 Q + k'''$$

$$R[fh' - (h + k)f'] \;=\; f''$$

$$R[gk' - (h + k)g'] \;=\; g''$$

$$h(0) = k(0) = h'(0) = k'(0) = h'(1) = k'(1) = 0$$

$$h(1) + k(1) = f(0) = g(0) = 1, \; f(1) = g(1) = 0$$

where β is the eccentricity of the elliptic base and f, g, h, k represent velocities and pressures in some coordinate system.

Let

$$v \;=\; \begin{pmatrix} f'\;(0) \\ g'\;(0) \\ h''\;(0) \\ k''\;(0) \\ Q \end{pmatrix} \;,\qquad F(v) \;=\; \begin{pmatrix} f(1) \\ g(1) \\ h(1) + k(1) - 1 \\ h'(1) \\ k'(1) \end{pmatrix} \;,\; \text{and}$$

$\rho_a(\lambda,v) = \lambda F(v) + (1 - \lambda)(v - a)$ be the homotopy used to solve $F(v) = 0$, which is equivalent to the two-point boundary value problem. This approach worked very well. An interesting result is that the most efficient direction in which to operate the slider is along its minor axis, i.e., sideways.

SQUEEZING OF A VISCOUS FLUID BETWEEN ELLIPTIC PLATES

The governing (nondimensional) equations are [45]:

$$f''' + K = S[2f' + \eta f'' + \tfrac{1}{2} f'f' - \tfrac{1}{2} f'' (f + g)]$$

$$g''' + \beta K = S[2g' + \eta g'' + \tfrac{1}{2} g'g' - \tfrac{1}{2} g'' (f + g)]$$

$$f(0) = g(0) = f'' (0) = g'' (0) = f'(1) = g'(1) = 0,$$

$$f(1) + g(1) = 2$$

where β is the eccentricity of the ellipses, S is a Reynolds
number, f and g describe the flow, and K is a constant·to be
determined. v, F(v), and ρ_a (λ,v) are defined analagously to the
elliptic porous slider problem. This problem displays extreme
sensitivity for S > 20, and very complicated behavior for S < 0.
Figure 4 shows the complicated geometry of the solution surfaces
for a particular set of the parameters (note the multiple solutions
and catastrophe at β = 1).

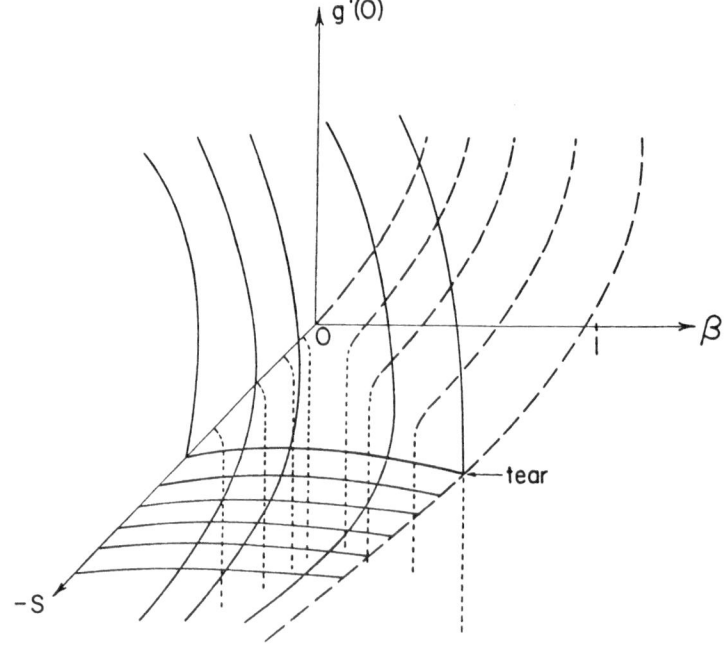

Figure 4. Solution surface for elliptic plates.

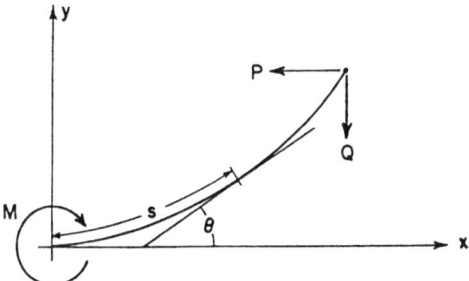

Figure 5.   Elastic rod.

ELASTIC ROD

Consider a thin incompressible elastic rod clamped at the origin and acted on by forces Q, P and torque M (see Figure 5). The governing (non-dimensional) equations are:

$$\frac{dx}{ds} = \cos \theta , \quad \frac{dy}{ds} = \sin \theta , \quad \frac{d\theta}{ds} = Qx - Py + M \tag{16}$$

$$x(0) = y(0) = \theta(0) = 0 \tag{17}$$

$$x(1) = a , \quad y(1) = b , \quad \theta(1) = c \tag{18}$$

The cantilever beam problem, which has a closed form solution in terms of elliptic integrals, is to find the position (a,b) of the tip of the rod given the forces $Q \neq 0$ and $P = 0$. Consider the inverse problem, where the a, b, c are specified, and Q, P, M are to be determined. For large c, c = 10π for example, the elastica is wound like a coil spring and its shape is extremely sensitive to small perturbations in Q, P, or M. For large deformations the problem (16-18) is ferociously nonlinear, and Newton and quasi-Newton methods generally fail [63].

The Chow-Yorke algorithm was completely successful on (16-18) using the homotopy map

$$\psi(d,\lambda,v) = \begin{pmatrix} x(1;v) - [\lambda a + (1 - \lambda)d_1] \\ y(1;v) - [\lambda b + (1 - \lambda)d_2] \\ \theta(1;v) - [\lambda c + (1 - \lambda)d_3] \end{pmatrix}$$

where $v = \begin{pmatrix} Q \\ P \\ M \end{pmatrix}$ and $x(s;v)$, $y(s;v)$, $\theta(s;v)$ are the

solution to the initial value problem (16-17).   In [63] numerous
approaches to this inverse elastica problem were considered, with
a homotopy method using the above homotopy map being the most
successful.   The homotopy

$$\rho_a(\lambda,v) = \lambda F(v) + (1 - \lambda)(v - a)$$

with $v = \begin{pmatrix} Q \\ P \\ M \end{pmatrix}$, $F(v) = \begin{pmatrix} +1 & 0 & 0 \\ 0 & +1 & 0 \\ 0 & 0 & \underline{+1} \end{pmatrix} \begin{pmatrix} x(1;v) - a \\ y(1;v) - b \\ \theta(1;v) - c \end{pmatrix}$

was unsuccessful on this problem for every sign combination.

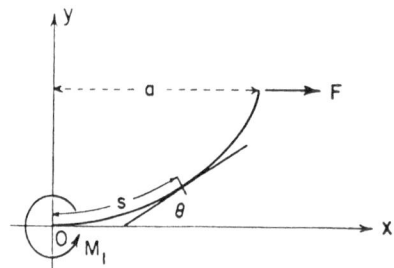

Figure 6.   Right half of a C-clamp.

C-CLAMP

Consider an elastic C-shaped clamp with natural curvature M_o as
shown in Figure 6.   The governing equations are similar to those
of the elastic rod, but the boundary conditions are different.
The equations are:

$$\frac{dx}{ds} = \cos\theta , \quad \frac{dy}{ds} = \sin\theta , \quad \frac{d\theta}{ds} = -Fy + M_1 + M_o ,$$

$$x(0) = y(0) = \theta(0) = 0 ,$$

$$x(1) = a , \quad \frac{d\theta}{ds}(1) = M_o \quad (M_o = \text{natural curvature}).$$

The solution details are similar to the elastic rod, and need not be repeated. For a complete discussion, see [47].

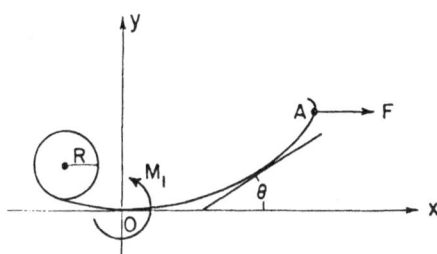

Figure 7. Negator clip.

NEGATOR CLIP

A related problem involves the negator clip or so-called "constant force spring". A spring with natural curvature M_o is wound into two coils of equal length (see Figure 7). It has been claimed [48] that the force exerted by the separated coils is independent of the separation of the coils. This is in fact true asymptotically, but the force F is a nonlinear function of lateral displacement X(L), where L is the arc length OA of the unwound spring, for moderate L/R ratios, where R is the natural radius of the spring. The governing equations are:

$$\frac{dx}{ds} = \cos\theta \, , \quad \frac{dy}{ds} = \sin\theta \, , \quad \frac{d\theta}{ds} = M_o - M_1 + Fy \, ,$$

$$x(0) = y(0) = \theta(0) = 0$$

$$\theta(L/R) = \pi/2 \, , \quad Fy(L/R) - M_1 = 0 \, ,$$

where M_o is the natural curvature, M_1 is the maximum moment occurring at the point of symmetry, and s is a nondimensional variable. For a complete discussion see [48].

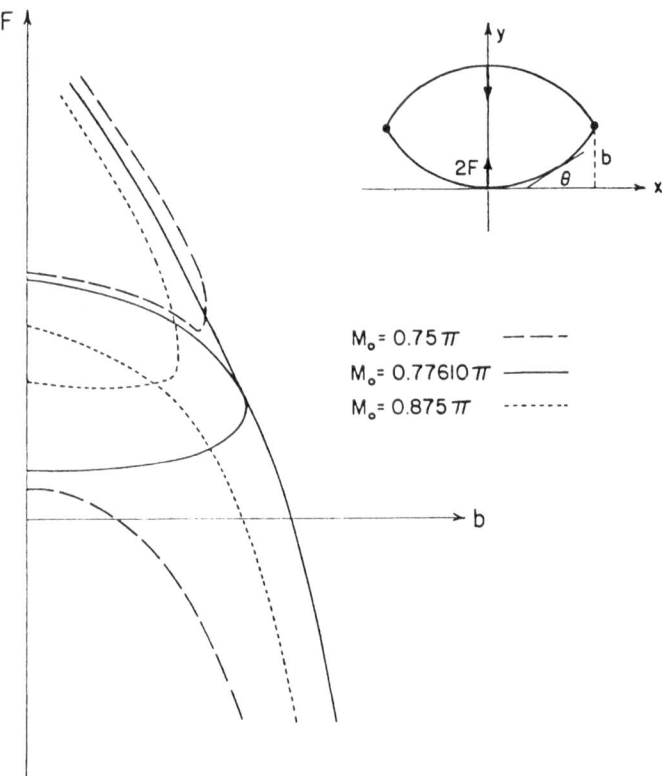

Figure 8.   Leaf spring and multiple solutions for
            several natural curvatures M_O.

LEAF SPRING

Another, but much more difficult, spring problem is the leaf spring
[64] (Figure 8). The governing equations

$$\frac{dx}{ds} = \cos \theta \, , \quad \frac{dy}{ds} = \sin \theta \, , \quad \frac{d\theta}{ds} = M_o - M + Fx$$

$$x(0) = y(0) = \theta(0) = 0$$

$$Fx(1) - M = 0 \, , \quad y(1) = b \qquad (M_o = \text{natural curvature})$$

are very similar to those for the negator clip, but there are
multiple solutions, turning points, and bifurcation points as
shown in Figure 8. The Chow-Yorke algorithm is not designed to
handle bifurcation points, and the bifurcation point shown in
Figure 8 was obtained by trial and error. The homotopy maps for
all of these elastica problems are similar to the elastic rod
homotopy. See [64] for a complete discussion of the leaf spring
problem.

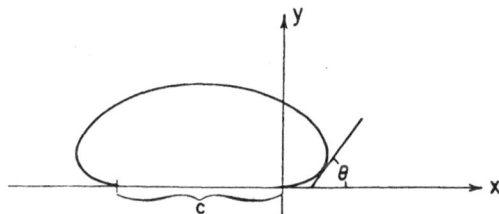

Figure 9. Fluid-filled cylindrical membrane container.

FLUID-FILLED CYLINDRICAL MEMBRANE CONTAINER

A rather different kind of elastica problem concerns a membrane
container filled with a fluid. Depending on the rigidity of the
container wall and the internal fluid pressure, the container sags
making contact with the ground (Figure 9). For low pressures and
rigidity, the cross-sectional shape is oblong and the container has
a small volume compared to a circular cylinder. For high internal
pressures or very rigid materal the shape is almost circular. The
interesting question is the trade off between pressure and volume,
since it is difficult and expensive to obtain high internal

pressures, yet low pressures waste container material since the
volume is comparatively small.   The (nondimensional) governing
equations are:

$$\frac{dx}{ds} = \cos\theta \;, \quad \frac{dy}{ds} = \sin\theta \;, \quad \frac{d\theta}{ds} = \frac{1}{\alpha}(\beta - y)$$

$$x(0) = y(0) = \theta(0) = 0$$

$$x(1 - c) = -c \;, \quad y(1 - c) = 0 \;, \quad \theta(1 - c) = 2\pi \;,$$

where β is a given constant, c is the unknown contact length,
and α is a parameter to be determined.   What makes this problem
different from the previous ones is that the interval of integration
$1 - c$ is unknown, and the boundary condition

$$x(1 - c) = -c$$

is difficult to handle.   Nevertheless, a straightforward homotopy
was successful [49].

HEAVY ELASTIC CYLINDER

Important construction problems in outer space and undersea involve
heavy elastic cylinders.   Depending on the rigidity of the elastic
wall material, the cylinder may collapse under its own weight.   There
are four distinct cases, governed by a nondimensional parameter
B   (see Figure 10).   Starting from a   perfect   cylinder (B = 0),
as   B   increases the point contact (Case 1) widens to a line
contact (Case 2) then the top sags until it touches the bottom for
a point-line contact (Case 3), then ultimately the top also makes
a line contact with the bottom.   The governing equations for all
four cases are

$$\frac{dx}{ds} = \cos\theta \;, \quad \frac{dy}{ds} = \sin\theta \;,$$

$$\frac{d^2\theta}{ds^2} = A\sin\theta + (C - Bs)\cos\theta .$$

For Case 1, C = B   and the boundary conditions are

$$x(0) = y(0) = \theta(0) = 0 \;,$$

$$x(1) = 0 \;, \quad \theta(1) = \pi .$$

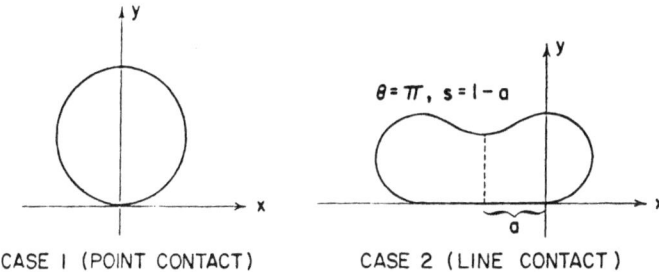

CASE I (POINT CONTACT)   CASE 2 (LINE CONTACT)

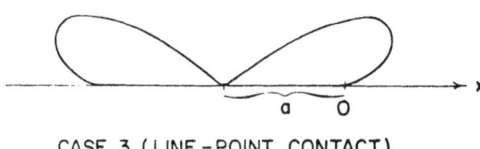

CASE 3 (LINE -POINT CONTACT)

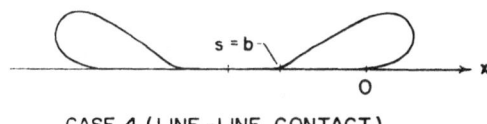

CASE 4 (LINE -LINE CONTACT)

Figure 10

For Case 2, $C = B(1 - a)$ and the boundary conditions are

$x(0) = y(0) = \theta(0) = \dot{\theta}(0) = 0$,

$x(1 - a) = -a$, $\theta(1 - a) = \pi$.

For Case 3, the boundary conditions are

$x(0) = y(0) = \theta(0) = \dot{\theta}(0) = 0$,

$x(1 - a) = -a$, $y(1 - a) = 0$, $\theta(1 - a) = \pi$.

For Case 4, the boundary conditions are

$x(0) = y(0) = \theta(0) = \dot{\theta}(0) = 0$,

$y(b) = 0$, $\dot{\theta}(b) = 0$.

For Cases 1 and 2, quasi-Newton methods are adequate and efficient if a good computer code is used. For Cases 3 and 4, where B is large, quasi-Newton methods are feasible but very expensive because of their small domain of practical application. If the starting point is too far away from the solution, quasi-Newton codes such as HYBRJ from Argonne's MINPACK fail to make progress toward the solution and give an error return [50]. The homotopy map

$\rho_a(\lambda, v) = \lambda F(v) + (1 - \lambda)(v - a)$,

where v consists of the appropriate initial conditions and parameters (depending on the case) and $F(v)$ is defined by shooting, works very well for large B [50]. This is a rare example of a problem on which quasi-Newton methods do not totally fail, and yet the homotopy algorithm is more efficient. Generally, quasi-Newton methods, when they work, are an order of magnitude more efficient than homotopy methods.

4. Conclusion

Differential geometry provides a solid theoretical foundation for
the Chow-Yorke algorithm [2,3], and homotopy maps producing globally
convergent algorithms have been constructed for a wide range of
problems. Perhaps the most spectacular successes have been for
Brouwer fixed points [51] and the nonlinear complementarity problem
[53]. The numerous engineering problems discussed here show that
homotopy methods are frequently successful on problems to which
the (known) theory is not applicable. The prospect of a globally
convergent algorithm, particularly on problems for which the best
quasi-Newton computer code [37] fails, makes homotopy methods
appealing and promising for future development.

On the negative side, the supporting differential geometry theory
requires at least C^2 smoothness, which means the Chow-Yorke
algorithm cannot handle directly, e.g., piecewise linear maps
(see [4], though). Also, developing a homotopy map whose zero
curves are bounded is very difficult, and, at present, an art.
Finally, homotopy methods are computationally expensive (at least
an order of magnitude worse than quasi-Newton methods), and there
is general agreement that they should only be used as a last resort.

References

1. L.R. Abrahamsson, H.B. Keller, and H.O. Kreiss, Difference
    approximations for singular perturbations of systems of
    ordinary differential equations, Numer. Math., 22(1974),
    pp. 367-391.
2. J.C. Alexander, The topological theory of an imbedding method,
    in Continuation Methods, H.G. Wacker, ed., Academic Press,
    New York, 1978, pp. 37-68.
3. J.C. Alexander and J.A. Yorke, The homotopy continuation method:
    numerically implementable topological procedures, Trans.
    Amer. Math. Soc., 242(1978), pp. 271-284.
4. J.C. Alexander, R.B. Kellogg, T.Y. Li, and J.A. Yorke, Piece-
    wise smooth continuation, Proc. NATO Advanced Research
    Institute on Homotopy Methods and Global Convergence,
    Sardegna, Sardinia, June 1981.
5. E. Allgower and K. Georg, Simplicial and continuation methods
    for approximating fixed points, SIAM Rev., 22(1980),
    pp. 28-85.
6. P. Boggs, The solution of nonlinear systems of equations by
    A-stable integration techniques, SIAM J. Numer. Anal.,
    8(1971), pp. 767-785.
7. P. Businger and G.H. Golub, Linear least squares solutions by
    Householder transformations, Numer. Math., 7(1965),
    pp. 269-279.
8. A. Charnes, C.B. Garcia, and C.E. Lemke, Constructive proofs
    of theorems relating to F(x) = y, with applications,
    Math. Programming, 12(1977), pp. 328-343.
9. S.N. Chow, J. Mallet-Paret, and J.A. Yorke, A homotopy method
    for locating all zeros of a system of polynomials, in
    Functional Differential Equations and Approximation of
    Fixed Points, H.O. Peitgen and H.O. Walther, eds.,
    Springer Verlag Lecture Notes in Math #730, New York, 1979,
    pp. 228-237.
10. S.N. Chow, J. Mallet-Paret, and J.A. Yorke, Finding zeros of
    maps: homotopy methods that are constructive with proba-
    bility one, Math. Comp., 32(1978), pp. 887-899.
11. G. Dahlquist, A. Bjorck, and N. Anderson, Numerical Methods,
    Prentice-Hall, Englewood Cliffs, N.J., 1974.
12. J. E. Dennis and J.J. Moré, Quasi-Newton methods-motivation
    and theory, SIAM Review, 19(1977), pp. 46-79.
13. F.W. Dorr, The numerical solution of singular perturbations
    of boundary value problems, SIAM J. Numer. Anal., 7(1970),
    pp. 281-313.
14. B.C. Eaves, Homotopies for computation of fixed points, Math.
    Programming, 3(1972), pp. 1-22.
15. B.C. Eaves and H. Scarf, The solution of systems of piecewise
    linear equations, Math. Operations Res., 1(1976),
    pp. 1-27.

16. B.C. Eaves and R. Saigal, Homotopies for computation of fixed points on unbounded regions, Math. Programming, 3(1972), pp. 225-237.
17. M.L. Fisher and F.J. Gould, A simplicial algorithm for the nonlinear complementarity problem, Math. Programming, 6(1974), pp. 281-300.
18. M.L. Fisher, F.J. Gould, and J.W. Tolle, A new simplicial approximation algorithm with restarts: Relations between convergence and labelling, Fixed Point Algorithms and Applications, S. Karamardian and C.B. Garcia, eds., Academic Press, New York, 1977, pp. 41-58.
19. C.B. Garcia, A global existence theorem for the equation Fx = y, Center Math. Studies Bus. Econ. Rep. 7527, Univ. of Chicago, Chicago, IL, 1975.
20. C.B. Garcia and F.J. Gould, Scalar labelings for homotopy paths, Math. Programming, 17(1979), pp. 184-197.
21. C.B. Garcia and F.J. Gould, A theorem on homotopy paths, Math. Operations Res., 3(1978), pp. 282-289.
22. C.B. Garcia and W.I. Zangwill, Determining all solutions to certain systems of nonlinear equations, Math. Operations Res., 4(1979), pp. 1-14.
23. C.B. Garcia and F.J. Gould, Relations between several path following algorithms and local and global Newton methods, SIAM Rev., 22(1980), pp. 263-274.
24. C.B. Garcia and T.Y. Li, On the number of solutions to polynomial systems of equations, MRC Rep. 1951, Univ. of Wisconsin, Madison, WI, April, 1979.
25. F.J. Gould and J.W. Tolle, A unified approach to complementarity in optimization, Discrete Math., 7(1974), pp. 225-271.
26. F.J. Gould and J.W. Tolle, An existence theorem for solutions to f(x) = 0, Math. Programming, 11(1976), pp 252-262.
27. H.B. Keller, Numerical Solution of Two-point Boundary Value Problems, SIAM, Philadelphia, 1976.
28. H.B. Keller, Numerical solution of bifurcation and nonlinear eigenvalue problems, in Applications of Bifurcation Theory, Academic Press, New York, 1977.
29. R.B. Kellogg, T.Y. Li, and J. Yorke, A constructive proof of the Brouwer fixed-point theorem and computational results, SIAM J. Numer. Anal., 13(1976), pp. 473-483.
30. R.W. Klopfenstein, Zeros of nonlinear functions, J. ACM, 8(1961), pp. 336-373.
31. M. Kubicek, Dependence of solutions of nonlinear systems on a parameter, ACM-TOMS, 2(1976), pp. 98-107.
32. T.Y. Li and J.A. Yorke, A simple reliable numerical algorithm for following homotopy paths, MRC Tech. Rep. 1984, Univ. of Wisconsin, Madison, 1979.
33. T.Y. Li and J.A. Yorke, Finding all the roots of polynomials by a homotopy method-numerical investigation, Dept. of Math., Michigan State Univ., East Lansing, MI, 1979.

34. R. Menzel and H. Schwetlick, Zur Lösung parameterabhängiger
    nichtlinearer Gleichungen mit singularen Jacobi-Matrizen,
    Numer. Math., 30(1978), pp. 65-79.

35. O. Merrill, Applications and extensions of an algorithm to
    compute fixed points of upper semicontinuous mappings,
    Doctoral thesis, I.O.E. Dept., University of Michigan,
    Ann Arbor, Michigan, 1972.

36. G. Meyer, On solving nonlinear equations with a one-parameter
    operator imbedding, SIAM J. Numer. Anal., 5(1968),
    pp. 739-752.

37. J.J. Moré, MINPACK documentation, Argonne National Lab.,
    Argonne, IL, 1979.

38. J.M. Ortega and W.C. Rheinboldt, Iterative solution of non-
    linear equations in several variables, Academic Press,
    New York, 1970.

39. R. Saigal, On the convergence rate of algorithms for solving
    equations that are based on methods of complementary pivoting,
    Math. Operations Res., 2(1977), pp. 108-124.

40. R. Saigal and M.J. Todd, Efficient acceleration techniques for
    fixed point algorithms, SIAM J. Numer. Anal., 15(1978),
    pp. 997-1007.

41. L.F. Shampine, H.A. Watts, and S.M. Davenport, Solving non-
    stiff ordinary differential equations - the state of the
    art, SIAM Review, 18(1976), pp. 376-411.

42. L.F. Shampine and M.K. Gordon, Computer Solution of Ordinary
    Differential Equations: The Initial Value Problem,
    W.H. Freeman, San Francisco, 1975.

43. S. Smale, Convergent process of price adjustment and global
    Newton methods, J. Math. Econom., 3(1976), pp. 107-120.

44. C.Y. Wang, The squeezing of a fluid between two plates,
    J. Appl. Mech., 43(1976), pp. 579-583.

45. C.Y. Wang and L.T. Watson, Squeezing of a viscous fluid between
    elliptic plates, Appl. Sci. Res., 35(1979), pp. 195-207.

46. C.Y. Wang and L.T. Watson, Viscous flow between rotating discs
    with injection on the porous disc, Z. Angew. Math. Phys.,
    30(1979), pp. 773-787.

47. C.Y. Wang and L.T. Watson, On the large deformations of
    C-shaped springs, Internat. J. Mech. Sci., 22(1980),
    pp. 395-400.

48. C.Y. Wang and L.T. Watson, Theory of the constant force spring,
    J. Appl. Mech., 47(1980), pp. 956-958.

49. C.Y. Wang and L.T. Watson, The fluid-filled cylindrical mem-
    brane container, J. Engrg. Math., 15(1981), pp. 81-88.

50. C.Y. Wang and L.T. Watson, Equilibrium of heavy elastic cylin-
    drical shells, J. Appl. Mech., to appear.

51. L.T. Watson, A globally convergent algorithm for computing
    fixed points of C^2 maps, Appl. Math. Comput., 5(1979),
    pp. 297-311.

52. L.T. Watson, Fixed points of C^2 maps, J. Comput. Appl. Math., 5(1979), pp. 131-140.

53. L.T. Watson, Solving the nonlinear complementarity problem by a homotopy method, SIAM J. Control Optimization, 17(1979), pp. 36-46.

54. L.T. Watson, T.Y. Li, and C.Y. Wang, Fluid dynamics of the elliptic porous slider, J. Appl. Mech., 45(1978), pp. 435-436.

55. L.T. Watson, An algorithm that is globally convergent with probability one for a class of nonlinear two-point boundary value problems, SIAM J. Numer. Anal., 16(1979), pp. 394-401.

56. L.T. Watson and D. Fenner, Chow-Yorke algorithm for fixed points or zeros of C^2 maps, ACM Trans. Math. Software, 6(1980), pp. 252-260.

57. L.T. Watson and W.H. Yang, Optimal design by a homotopy method, Applicable Anal., 10(1980), pp. 275-284.

58. L.T. Watson, Computational experience with the Chow-Yorke algorithm, Math. Programming, 19(1980), pp. 92-101.

59. L.T. Watson and C.Y. Wang, Deceleration of a rotating disc in a viscous fluid, Phys. Fluids, 22(1979), pp. 2267-2269.

60. L.T. Watson, Numerical study of porous channel flow in a rotating system by a homotopy method, J. Comput. Appl. Math., 7(1981), pp. 21-26.

61. L.T. Watson, Solving finite difference approximations to nonlinear two-point boundary value problems by a homotopy method, SIAM J. Sci. Stat. Comput., 1(1980), pp. 467-480.

62. L.T. Watson and W.H. Yang, Methods for optimal engineering design problems based on globally converegent methods, Computers & Structures, 13(1981), pp. 115-119.

63. L.T. Watson and C.Y. Wang, A homotopy method applied to elastica problems, Internat. J. Solids Structures, 17(1981), pp. 29-37.

64. L.T. Watson and C.Y. Wang, The circular leaf spring, Acta Mechanica, to appear.

65. L.T. Watson and C.Y. Wang, Hanging an elastic ring, Internat. J. Mech. Sci., 23(1981), pp. 161-168.

66. L.T. Watson, S.M. Holzer, and M.C. Hansen, Tracking nonlinear equilibrium paths by a homotopy method, Computers & Structures, to appear.

AVAILABILITY OF COMPUTER CODES FOR PIECEWISE-LINEAR

AND DIFFERENTIABLE HOMOTOPY METHODS

During discussions taking place at the conference it became apparent that information about available implementations of homotopy methods was not generally known and that a list of such codes would provide a valuable service to both researchers and potential users. The next few pages give some details of available programs provided by their authors. We do not claim that the list is complete and we cannot vouch for the accuracy of all the information. Nevertheless, we hope that researchers and potential users will gain some appreciation for the range of codes available and will be encouraged to contact their authors.

BOONE

General Description

BOONE is used to compare the performance of various predictor-corrector methods for following homotopy paths. The predictors implemented use Hermite interpolation, and thus include the Euler predictor as well as predictors based on Lagrange and osculatory interpolation. The corrector implemented is the pure Newton method. The homotopies implemented include the fixed point homotopy (Levenberg-Marquardt) and the Newton homotopy. The predictor is chosen at run time by choice of certain input data. The corrector and homotopy are chosen at load time by selection of subroutine object deck.

Type of Problem and Data Requirements

The code facilitates both zero finding (for functions from R^n to R^n) and more general path following (in the manifold of zeroes of a homotopy from R^{n+1} to R^n). The user who wants a zero of a function must provide (1) subroutines to evaluate the function and its first partial derivatives, (2) the values of certain parameters (e.g., error tolerances), and (3) a starting point. The multi-point Lagrange and osculatory predictors are most suited for close path following. The Euler predictor is best on other types of problems.

Program Language and Length

BOONE is written in ANSI standard FORTRAN 77 and is about 1200 lines long.

Documentation and Availability

Documentation and an export version of the code are being prepared and will be made available to anyone supplying a tape. Contact the author for more current information.

Author

W.F. Griffeth, College of Management, Georgia Institute of Technology, Atlanta, GA 30332, USA.

FIXPT

General Description

FIXPT implements the Chow-Yorke algorithm by a homotopy to the trivial map. It utilizes L. F. Shampine's sophisticated ODE code and numerically stable matrix factorizations. It is very reliable and robust, but expensive.

Type of Problem and Data Requirements

FIXPT is designed to compute Brouwer fixed points or zeros of C^2 maps from E^n to E^n. There are clearly marked dimension statements in two subroutines limiting n to 100. These dimension statements can be changed and the linear algebra subroutines can be modified to handle large sparse problems. The user must supply two subroutines to evaluate the function and its Jacobian matrix.

Program Language and Length

FIXPT is written in portable FORTRAN (verified by the PFORT compiler). Two subroutines contain machine dependent constants, for which appropriate DATA statements must be chosed, as explained in the listing. The length is 1155 lines.

Documentation and Availability

The algorithm is documented in

L.T. Watson and D. Fenner, "Chow-Yorke algorithm for fixed points or zeros of C^2 maps", ACM Trans. Math. Software, 6 (1980) 252-260.

A complete listing is in "Collected Algorithms from ACM", and the code (cards or tape) may be obtained from the ACM Algorithms Distribution Service (see the journal for an order form).

Authors

Layne T. Watson, Department of Computer Science, Virginia Polytechnic Institute & State University, Blacksburg, VA 24061 USA

Dan Fenner, Johns Hopkins Applied Physics Laboratory, Johns Hopkins Road, Laurel, MD 20810 USA

FIXPT 1

General Description

FIXPT 1 is an implementation of the Restart algorithm of Merrill, and is available in two versions implementing two different triangulations K and H.

Type of Problem and Data Requirements:

The code is designed for a location of a fixed point of a continuous or an upper semi-continuous point to set mapping from R^n into itself. It can solve a problem of up to 98 variables. The user must supply a subroutine to evaluate the function and possibly a subroutine to read the data needed for this evaluation.

Program Language and Length:

The language used is FORTRAN IV, and the code can be used on any IBM machine without modifications. Otherwise, some changes may be needed. The program has about 360 lines of code.

Documentation and Availability:

The documentation and source code are available at a small fee, required to cover mailing and duplication. Contact the author.

Author

R. Saigal, Department of Industrial Engineering, Northwestern University, Evanston, Illinois 60203.

SUBROUTINE FIXPT

General Description

Subroutine FIXPT is an implementation of a piecewise linear homotopy algorithm based on the continuous deformation method implemented on the triangulation J_3. It achieves quadratic convergence for differentiable functions, and at the option of the user, can be hybridized with discrete Newton steps, to increase the efficiency of convergence. This subroutine can also exploit the resulting savings when the mappings are separable.

Type of Problem and Data Requirements

The subroutine is designed for computing a fixed point of continuous or upper semi-continuous point to set mappings from R^n into itself. It is dimensioned to solve a problem of up to 25 variables, but this can be readily increased by changing the dimension statements. The user must supply a subroutine to evaluate the functions, and possible routine to read the data needed in the evaluation of function. For more advanced users, by providing a subroutine, the printout of the program can be changed. This program is a subroutine, and parameters needed for initiating the program are provided by the calling statement. The user can avoid providing these parameters, and use the defaults provided in the subroutine. Thus, the user must also provide a main program.

Program Language and Length

Subroutine FIXPT is programmed in IBM Fortran IV, and may need some modifications for other machines. The source code provided is suitable for a CDC machine.

Documentation & Availability

The users manual Efficient Algorithm for Computing Fixed Points when Mappings may be separable - A Computer Program and Users Manual (update November 1970) and the source code are available. There is a small charge to cover duplication and postage. Contact the author.

Author

R. Saigal, Department of Industrial Engineering, Northwestern University, Evanston, Illinois 60203.

PLALGO

General Description

PLALGO is an implementation of several piecewise-linear homotopy algorithms, using a numerically stable factorization of the basis and discrete Newton acceleration.

Type of Problem and Data Requirements

The code is designed for location of a zero of a continuous function or upper semi-continuous point-to-set mapping from R^n to itself, using a homotopy from a linear function. It is dimensioned for $n \leq 50$, though this is very easy to increase. The user must supply a subroutine to evaluate the function or mapping and various parameters, most of which can be set automatically.

Program Language and Length

PLALGO is written in portable FORTRAN, with no machine-dependent constants; it contains about 2500 lines including extensive comments.

Documentation and Availability

Documentation is available. If demand is not excessive, the code will be sent to anyone supplying a tape; mailing costs may be requested. Contact the author.

Author

M.J. Todd, School of Operations Research and Industrial Engineering, Cornell University, Ithaca, NY 14853, USA.

SCOUT - Simplicial Continuation Utilities

General Description

SCOUT is a program designed to trace the solutions of nonlinear eigenvalue or bifurcation problems with one or two parameters, i. e. the zeroes of a mapping $F : R^n \times R^2 \to R^n$. SCOUT employs a piecewise linear algorithm using K_1 as a triangulation and updating the inverse of the labeling matrix. Routines for the purposes of generating a start, mesh refinement, acceleration of the pivoting scheme, finding all branches at bifurcation points, and Newton-like local iteration are included.

Operation

The user has to supply a subroutine to evaluate the map F . Moreover, the user interactively communicates with the program in order to initiate actions such as mesh refinements, perturbations, or acceleration and to control input and output data flow, e. g. creation of files containing data dumps or plot data. During program execution a help facility enables the user to obtain information about valid commands and their parameters.

Program Language and Length

SCOUT is written in portable FORTRAN except for one short subroutine written in FORTRAN77. It contains about 900 lines of executable statements.

Documentation and Availability

Some documentation is available. The code will be sent to anyone supplying a tape.

Authors

Hartmut Jürgens and Dietmar Saupe, Forschungs-schwerpunkt Dynamische Systeme, Universität Bremen, Bibliothekstraße, 2800 Bremen-33, West Germany.

Remark

A new version of SCOUT utilizing a predictor corrector method, thus increasing efficiency, will be made available within 1981.

INDEX

317